OF APES AND ANCESTORS:
EVOLUTION, CHRISTIANITY, AND THE OXFORD DEBATE

IAN HESKETH

Of Apes and Ancestors

Evolution, Christianity, and the Oxford Debate

UNIVERSITY OF TORONTO PRESS
Toronto Buffalo London

Toronto Buffalo London
www.utppublishing.com
Printed in Canada

ISBN 978-0-8020-9284-7

Printed on acid-free paper

Library and Archives Canada Cataloguing in Publication

Hesketh, Ian, 1975–

Of apes and ancestors : evolution, Christianity, and the Oxford debate / Ian Hesketh.

Includes bibliographical references and index.
ISBN 978-0-8020-9284-7

1. Huxley, Thomas Henry, 1825–1895. 2. Wilberforce, Samuel, 1805–1873. 3. Darwin, Charles, 1809–1882. 4. Evolution (Biology) – Great Britain – History – 19th century. 5. Religion and science – Great Britain – History – 19th century. 6. Great Britain – Intellectual life – 19th century. I. Title.

BL263.H48 2009 213 C2009-903055-1

University of Toronto Press acknowledges the financial assistance to its publishing program of the Canada Council for the Arts and the Ontario Arts Council.

University of Toronto Press acknowledges the financial support for its publishing activities of the Government of Canada through the Book Publishing Industry Development Program (BPIDP).

Contents

Acknowledgments

This book has amassed a debt far out of proportion to its small size. Allan Hutchinson initially approached me about helping him with the research for a project on the 'Oxford debate,' a project for which I quickly became co-author and then sole author. Much of my thinking on the subject was developed out of conversations with Allan about our proposed joint project, and I owe him an immense debt of gratitude for encouraging me to write it on my own. I also gratefully acknowledge a Social Science and Humanities Research of Canada Postdoctoral Fellowship, which provided me with the time and funds necessary to research and write this book. I also wish to thank Virgil Duff, the executive editor at the University of Toronto Press, for believing in this project from its early stages and for having the manuscript reviewed so quickly. The three anonymous referees also provided very valuable suggestions and criticisms that have greatly improved the book. I must also thank Matthew Kudelka for doing such a thorough job copy editing the manuscript.

I should like to thank the staff of the Koerner, Barber, and Okanagan Libraries at UBC for tracking down various sources for me over the past few years. I would like to thank Hilary McEwan at the Special Collections of the Imperial College London Library for guiding me through the Huxley collection. Colin Harris arranged for my viewing of Wilberforce's papers at the Bodleian Library at Oxford University. I also must thank the staff at the Oxford University Museum of Natural History for showing me where the debate took place as well as sharing with me the museum's resources on the debate itself.

I can't let this opportunity pass without thanking my parents and sister for providing me with their limitless support over the years while I pursued my scholarly interests. I know that my work has not always been interesting to them, but they have done a wonderful job pretending it is, and that means so much to me. I know that they had much less trouble feigning their interest in this project. My partner Cleo kept me fed and clothed throughout the research and writing of this book, and it is impossible for me to weight the value that her friendship and love has on my work.

I wish it was possible to thank all of the professors I've had over the years who have influenced my interest in history. For this project in particular, I have been influenced and assisted in different but countless ways by James Hull, Bernard Lightman, and Eric Nellis. It was while a student in James's class on Modern Science in 1998 that I became interested in all things Darwin. I can still remember a lecture he gave on 'Darwin's God' that opened my eyes to Darwin's subtle but effective rhetorical strategies. He was kind enough to read my entire manuscript, and he provided excellent suggestions, most notably about recasting the first chapter to be about Darwin's 'historical sketch.' I was also lucky enough to have recently heard a new version of James's lecture on 'Darwin's God,' this time given to my own class. Bernie's lectures and immense work on Victorian science have been a wonderful resource and inspiration for my own foray into the vast field. He, too, read through this entire manuscript and proved to be a fount of information regarding secondary sources as well as the intricate nuances of Victorian science and religion and the current debate surrounding the New Atheists. Eric was the first history professor I had as an undergraduate student and his lectures were a thing of legend. I was fortunate to find myself in an office next to his during my postdoc at UBC, and he gave me much encouragement when the project was in its early stages. We met several times to discuss the manuscript, of which he read many versions, and he provided indispensable suggestions about the structure of the narrative and about the writing style itself. He convinced me of the project's value, and I couldn't have finished the manuscript without his help.

OF APES AND ANCESTORS:
EVOLUTION, CHRISTIANITY, AND THE OXFORD DEBATE

Introduction

> I am indeed most thoroughly contented with [the] progress of opinion [on the origin of species]. From all that I hear from several quarters, it seems that [the] Oxford [debate] did the subject great good.—It is of enormous importance the showing the world that a few first-rate men are not afraid of expressing their opinion. I see daily more and more plainly that my unaided Book would have done *absolutely* nothing.
>
> Charles Darwin to Thomas Henry Huxley, 20 July 1860

Charles Darwin's *Origin of Species*, published in 1859, would have made little impact had it not been defended by important scientific allies who were willing to give shape and meaning to its contents, putting their own reputations on the line in doing so. Darwin was himself aware of this fact, on 20 July 1860 writing to his most outspoken defender, Thomas Henry Huxley (1825–1895), that 'unaided' his 'Book would have done *absolutely* nothing.' Darwin was especially pleased with a public debate that had taken place at Oxford University a few weeks earlier at the annual meeting of the British Association for the Advancement of Science (BAAS). The debate itself had done little to advance the subject, but the stories and publicity produced by the debate generated much interest in Darwinian evolution while providing a narrative arc for Darwin's defenders. Facts alone would not be enough for evolution to make the jump from a heterodox and heretical theory of species development to one of scientific orthodoxy. History itself would need to be on the Darwinians' side.

Indeed, gaining scientific consensus for evolution by natural selection was an uphill battle particularly because the theory seemed to challenge the well-established tradition of natural history known in Britain as natural theology. William Paley's (1743–1805) widely read and influential work of the same name, published in 1802, taught that science could be used to understand and appreciate God's creations while providing direct evidence for His existence, wisdom, and goodness. A series of publications known as the Bridgewater Treatises, written by Britain's leading scientific thinkers in the 1830s, codified natural theology as a particularly dominant form of natural history. Darwinian evolution was suspect under the terms of this tradition because it excluded a Creator from the day-to-day workings of evolution, while appearing to posit a natural cause for organic origins – origins that, being a first cause, were therefore beyond the reach of scientific rationality.[1] The science of Paley and the Bridgewater Treatises included the supernatural and the natural: miracles when it came to first causes (such as organic origins), scientific laws when it came to second ones.

In the first half of the nineteenth century, however, a series of geological and archaeological discoveries directly challenged the Christian narrative of the earth's history found in the Old Testament's book of Genesis. As a result of these discoveries, the earth came to be understood as much older than previously thought. Moreover, the appearance of prehistoric human remains thrust man deep into the past, well before Adam and Eve supposedly tasted the forbidden fruits of knowledge. First in Germany and soon after in Britain, biblical criticism became a popular discipline of study, putting the Bible under intense scrutiny in much the same way that historians were beginning to examine historical documents from newly opened state archives. Meanwhile, the belief that the earth's species were not individually created by a divine act, but rather were products of a long process of evolution, began to gain support as a rival narrative to that of the Genesis account.

Evolution had baggage of its own, however. While it certainly explained many recent observations in several scientific disciplines, it could not help but undermine many of the core beliefs of a predominantly Anglican society – beliefs such as those in Divine Providence, the afterlife, and salvation. It even threatened the lessons in day-to-day morality taught by the Ten Commandments. And perhaps most important, it brought into question the special place that humans believed they occupied on the earth, at the top of a great chain of being – a chain that now threatened to become, in Darwin's metaphor, a tree on which humans occupied merely one branch among many.

Darwin was adamant, however, that in accepting evolution one need not throw out the baby with the bathwater, and indeed, many converts to evolution continued to hold traditional Anglican beliefs. But the contradiction in doing so was obvious to many others, and it was not uncommon for someone who accepted Darwinian evolution to experience a profound crisis of faith. One such figure was the literary critic and father of Virginia Woolf, Leslie Stephen, who reminisced late in his life about the immeasurable intellectual changes of his youth, changes that 'shocked' the generation growing into maturity in the 1850s.[2] Stephen, who had been brought up in a strict Anglican household, contemplated suicide once he grasped the social and moral dimensions of Darwinian evolution, a theory he accepted as scientific fact. Stephen's deep internal struggle against his Anglican upbringing – a struggle engendered in part by a belief in the theory of evolution – has become a familiar trope in the historian's conception of Victorian unbelief.[3] Indeed, whether or not Darwinian evolution was good science, both converts and critics agreed that it threatened to transform how humans understood themselves and how they saw the world around them. Debates about Darwinian evolution certainly centred around its scientific merits; but also at issue was what constituted the proper subject matter for a natural science, as well as who had the authority to make knowledge claims about that subject matter. The nature of science, the future of religion, and questions of cultural authority were all at stake.

This would become clear at the Oxford debate of 1860, the debate that Darwin believed did so much good for his book even though he was not present to witness or take part in the proceedings. It must be noted that this debate was not on the agenda during the discussion period of one of many sessions that took place throughout the week-long British Association meeting. This session, however, happened to focus on 'the Views of Mr. Darwin,' and there began a heated debate. The Bishop of Oxford, Samuel Wilberforce, was in the audience, and he used the discussion period as a pulpit to criticize the *Origin*. Many others spoke their mind, most notably Huxley and the botanist Joseph Hooker (1817–1911), both of whom defended Darwin's evolutionary hypothesis. Some might call this skirmish a minor incident, and in many ways it was. However, accounts of the debate spread, and this minor incident became a set piece of Victorian cultural lore.

The Bishop of Oxford, 'Soapy' Samuel Wilberforce, sought to 'smash' the materialism of Darwinism through ridicule and sarcasm, openly asking an adherent of evolution on which side of his family he traced his ancestry from the apes. Barely able to throw his voice over the chorus of laughter and applause, Thomas 'Darwin's Bulldog' Huxley rose to respond in kind, suggesting that it was far better to be descended from a primitive species than from a man who would use religious sophistry to obscure the truth. The headlines quickly read: SCIENTIST WOULD RATHER BE RELATED TO APE THAN BISHOP.[4] Or so the story goes.

When Darwin wrote to Huxley that, from what he heard, the Oxford debate did the 'subject great good,' he was referring to the buzz it had created rather than to any scientific point that had perhaps been won. He was also pleased about the narrative that was developing around the debate – that his supporters had refused to back down in the face of clerical attack. 'It is of enormous importance' Darwin wrote, 'showing the world that a few first-rate men are not afraid of expressing their opinion.' This suggests that the debate was not simply about evolution: it was also a battle between science and Christianity, with the former refusing to cower to the latter's power and authority. This, as it turned out, was the first draft of history, rewritten but not revised throughout the nineteenth and much of the twentieth centuries: the Oxford debate as an open battle between science and Christianity.

History has proven to be a powerful tool when it comes to spinning understandings of science that serve contemporary interests. For example, historians of science during the Enlightenment invented the historical period known as the Scientific Revolution to break the science of the Middle Ages from that of the early modern period; in doing so, they were directing a not so subtle criticism at the remnant (institutional) authority of the Church and the (intellectual) authority of religion in the eighteenth century. The historicization of Galileo's troubles is a perfect example of this: it was long supposed that his astronomical observations in support of a heliocentric cosmology were suppressed by the Catholic Church because it feared the impact his findings would have on its authority. We now know that this affair was more complicated – that Galileo openly ridiculed the Pope and that he himself was the product of an Italian system of courtly patronage that led to both his success and his eventual suppression.[5]

This melodramatic battle between science and Christianity was lent further credence in the twentieth century. Berthold Brecht's play *Galileo* (1937–39) reinforced the Enlightenment's simplistic interpretation of the Galileo Affair – that his house arrest was an issue of scientific rationality confronting the irrational dogmatism of the Catholic faith. Similarly, Jerome Lawrence and Robert E. Lee's play *Inherit the Wind* (1955) dramatized the Scopes Monkey Trial (1925) – a trial that involved the prosecution of John Scopes, who had defied the Butler Act in Dayton, Tennessee, by teaching the theory of evolution to his high school students. In a veiled attempt to criticize the anti-communist investigations of the House Un-American Activities Committee, the play turned the trial into a story of heroes and villains, with the prosecutor William Jennings Bryan sounding like a hate-filled Senator Joseph McCarthy. Indeed, the links the authors drew between Christianity and the intellectual and political suppression associated with the anti-communist hysteria of the 1950s were not subtle, even though the actual trial was essentially a put-up job by Dayton officials and lawyers looking to gain some publicity and an economic boost for their city.[6]

The nineteenth century had not been immune to this type of mythologizing. The Enlightenment's narrative of rational science fighting to overcome irrational Christianity was born anew in the nineteenth century as historians reached into the past and found science suffering at the hands of Christianity in the same way that evolution was supposedly suffering at the hands of natural theology. Narratives of the Oxford debate, which were a product of this historical spin doctoring, have survived right up to the present. Indeed, just as Creationists in the United States have engaged in a pitched battle for Creationism to be taught alongside evolution in schools, evolutionary supporters and religion baiters such as Christopher Hitchens and Philip Kitcher – who have simplified the history of evolution and its relationship with Christianity well past the point of distortion – have reached into their historical arsenals to pull out the Oxford debate. Hitchens has described the Oxford debate as a 'tipping point' in the battle between evolution and Christianity where 'in front of a large audience, Huxley cleaned Wilberforce's clock, ate his lunch, used him as a mop for the floor, and all that.'[7] Yet there is little evidence in the historical record to support such a one-sided victory for either man. (In the same article Hitchens suggests that *Inherit the Wind* gave 'a fair summary' of the Scopes Trial.) Yet Hitchens, who admits that he learned the story of the Oxford debate in school, continues to tell it as if it was a bare fact. His belief in the truth of evolution has caused him to degrade the early anti-Darwinians, who it must be noted did not have the benefit of the past fifteen decades of scientific development – decades during which evolutionary theory has generated a powerful and ever-expanding research program. Creationists in the twenty-first century are not the same as proponents of special creation in the mid-nineteenth, many of whom later became supporters of evolution.

Darwin's book, on publication, did generate a sense of urgency because it seemed to be more than a mere scientific theory. A close examination of the Oxford debate and of the key participants' personalities and social contexts will make it clear that the initial conflict surrounding Darwinian evolution extended far beyond the boundaries of science. At the same time, the opposition to Darwin's theory was not solely religious. Darwinian evolution injected itself into a complex web of cultural debates that by that time was raging among geologists, naturalists, biologists, paleontologists, morphologists, biblical scholars, various religious sects, politicians, clergymen, historians, industrialists, merchants, aristocrats, socialists, and Chartists. Mid-Victorian Britain was a rapidly transforming society, one in which 'shock cities' – new industrial communities – were springing up within mere decades, in which new political actors were arriving on the scene and a capitalist economy was expanding at the same rapid rate as an urban underclass. The printing press, the telegraph, and the steam engine were collapsing space and time, with the result that new and often frightening ideas were being transported to the far edges of the Empire within days of being printed. Most significant, perhaps, was not the fact of Victorian Britain's rapidly changing cultural and social landscape but rather Victorians' sharp awareness of this change. 'Every age may be called an age of transition,' argued Edward Bulwer as early as 1833, 'but in our age the transition is *visible*.'[8] Novelist, historian, and clergyman Charles Kingsley wrote in 1855 of the immense transition of the age, that he felt as if standing 'on a cliff which is crumbling beneath one, and falling piecemeal into the dark sea.' The historian John Morley similarly argued that 'those who dwell in the tower of ancient faiths look about them in constant apprehension, misgiving, and wonder, with the hurried uneasy mien of people living amid earthquakes. The air seems full of missiles, and all is doubt, hesitation, and shivering expectancy.'[9] Perhaps this is what Marx, Darwin's admiring contemporary, had in mind after he visited Manchester when he and Engels wrote that 'all that is solid melts into air' in describing the impact of bourgeois capitalism in the *Communist Manifesto*.[10] New ideas had a palpable weight to them, and with the rise of the mass media there was little chance of escaping or even ignoring their power.

Darwinian evolution would become one such idea and a powerful exemplar of a swiftly changing society representing a host of converging realities.

The Oxford debate took place at a time when the debates over Darwin's book were reaching a fever pitch, and it came to symbolize both the opposition to and the defence of Darwin's evolutionary hypothesis. Like most famous historical events, however, the Oxford debate is as much myth as it is reality. This book is an attempt to separate the facts from the fiction. At the same time, we should not deny the historical significance either of the debate itself or of the myths that surround it. It was a momentous occasion, one during which the most important scientific and religious authorities of the day gathered to discuss what would become by far the most important scientific work of the century. However, it is hoped that this book will diminish a particular meaning of the debate while highlighting other meanings that are less obvious but up until now have been downplayed. Typically, the debate is presented as an example of the inevitable battle between science and Christianity, as an attempt by organized religion to hold back the tide of scientific discovery and authority. The picture presented here will be much more nuanced than this tired image of science versus religion. By unpacking the Oxford debate through an examination of its dramatis personae, this book will present a fuller picture of the event, in the course of which personal and professional conflicts blended with religious and scientific ones.

One needs to be careful when generalizing about the historical relationship between science and religion. But it is worth noting here that the rise of evolution clearly created new boundaries for both religion and science in Britain. Darwin expanded the scope of scientific analysis to incorporate the history of species, man included. This had long been the terrain of religion and of the compromised science of natural theology. The Oxford debate was part of the breaking down of this compromise: a younger generation of scientists saw in the *Origin* a chance to expand the scale and the scope of secular science, to free science from the shackles of Anglican dictates. In this sense the Oxford debate was a central event in the professionalization of science. But it was also the product of personal struggles and jealousies that had little to do with the big picture of science and religion.

This study of the Oxford debate, then, proceeds along two interrelated paths. The debate will be examined from the perspective of each of the key participants, providing insight into the intellectual climate of mid-Victorian Britain. It will also be necessary to consider the ways in which the debate has been mythologized as a great clash between science and Christianity. Indeed, the 'victors' of the debate wrote the initial histories, and the debate itself provided a narrative arc for their continued cause of a secularized science. It became a Galileo Affair for the nineteenth century. Because of this, any attempt to reconstruct the Oxford debate must also reconstruct the cultural memory of the debate itself – in other words, it must examine how the debate has been written about and remembered. From this perspective the debate illuminates the relationship between science and history, science and Christianity, and the dynamics among all three. But first we must turn to the context of Darwin's revolution, to the intellectual crisis of mid-Victorian Britain, and to the people who sought to shape Darwin's legacy. With this in mind, the first four chapters will introduce the reader to the debate's dramatis personae, beginning with Darwin, who was present in spirit even though he did not physically attend. The final two chapters will discuss the debate and its historical representations.

1 Charles Darwin: Historian of Natural History

The task [of writing a historical preface] would have been not a little difficult, and belongs rather to the Historian of Science than to me.

Charles Darwin to Baden Powell, 18 January 1860

I have resolved to publish a little sketch of the progress of opinion on the change of species.

Charles Darwin to Joseph Hooker, 31 January 1860

Sometime between 18 January and 31 January 1860, Darwin decided to write 'An Historical Sketch,' a brief historiography of the idea of transmutation that would act as a preface to the American and German editions of the *Origin* published later that year and to all future English editions beginning with the third in 1861. The publishing history of the *Origin* is a fascinating one, in that Darwin made often subtle but also significant changes throughout the six English editions published between 1859 and 1872.[1] Darwin answered his critics not by writing responses to the many periodicals and newspapers where reviews of the *Origin* appeared or by debating his foes in the public sphere of scientific and learned societies. Instead, he responded by continually revising the *Origin* in order to take into account new evidence but also new problems exposed by critics and friends alike. The various changes made to the six editions of the *Origin* can be read as Darwin's private responses to a public debate that was raging throughout Britain about his book in particular and evolution in general. The addition of the historical sketch was Darwin's attempt to silence critics who argued that he had not given proper due to his evolutionary precursors. It also allowed Darwin to make clear the originality and priority of his theory while distancing himself from the more radical evolutionary theories of the past. It seems that from the very beginning, historical spin doctoring was being used to condition the reception of the *Origin*.

Before discussing Darwin's historical sketch, certain aspects of his biography must be considered that have a bearing on the way in which he wanted his theory to be received and remembered. For instance, he disliked public conflicts, preferring to let his writing (and his friends) do the talking for him. He therefore tended to avoid public meetings where his book was likely to take centre stage. Darwin's absence from the annual meeting of the British Association for the Advancement of Science (BAAS) in late June of 1860 is a case in point. His *Origin of Species*, published less than a year earlier, had sold out immediately and had quickly been reprinted. It was fast becoming the must-read book of the literary season and was generating wide debate about the possible transmutation of men from apes – a topic never actually discussed in the book, which supposedly supported the claim. Reviews of the *Origin* were pouring in by this time, some glowingly positive but many more ruthlessly critical of the man who appeared to be sullying Christian theology by linking humankind to other animals. Darwin and his close scientific friends viewed the BAAS as an enclave of old-school natural theologians, gentleman naturalists who believed that the study of nature and the study of God went hand in hand – in other words, BAAS was an association not likely to be friendly to a work that described endless changes in species, changes that were not caused by God's invisible hand. But at the same time, BAAS was really the only place in Britain where scientific minds could meet on a large public scale. Founded in 1831, it served as a venue for scientists to meet, socialize, and present research in progress. It also had a propagandistic function, in that its annual meetings were always held in a provincial town and were open to the public. It was hoped this would spread the interests of science throughout the country.[2] This was precisely the type of meeting that Darwin tended to avoid, and he thought it best to use his well-worn excuse of illness to justify his absence.

Instead of meeting Thomas Henry Huxley and Joseph Hooker in Oxford to attend readings of various papers on all things scientific, Darwin travelled to Sudbrook Park, Richmond, for a few days' rest and relaxation, where he could also receive treatment for his constant stomach churning, flatulence, and retching, a treatment Darwin called his water cure.[3] Darwin was, as Adrian Desmond and James Moore describe him, a tormented evolutionist.[4] He was not the type of person to seek out popularity or conflict, and he no doubt would have been more than happy to live out his days in the village of Downe as an uncontroversial naturalist.[5] He was, however, too controversial a figure to hide. While on his historic HMS *Beagle* voyage (1831–36) he had been 'much struck,' as he put it in the *Origin*, 'with certain facts in the distribution of the organic beings inhabiting South America, and in the geological relations of the present to the past inhabitants of that continent.' When he returned home and examined his specimens he realized that he might be able to 'throw some light on the origin of species – that mystery of mysteries' – and he set about 'patiently accumulating and reflecting on all sorts of facts which could possibly have any bearing on it.'[6] Darwin was very careful when presenting his theory to show that it was the result of patient and meticulous observation, that he had not arrived at an evolutionary view of species lightly or without proper scientific objectivity. However useful this was as a strategy of self-presentation, Darwin was not lying:[7] he *did* collect and observe as many data on the question of species as he could, primarily from evidence related to his *Beagle* voyage but also at home. Indeed, he wrote letter after letter, constructing an immense network of correspondence extending to the far reaches of the globe, in search of answers to his varied and multiplying questions with regard to species and their histories.

A young naturalist named Alfred Russel Wallace was one of Darwin's correspondents. Darwin had earlier written to Wallace – who at the time was a little-known naturalist about to journey to the Malay Archipelago (1854) – to inquire into the possibility of being sent some specimens.[8] Darwin's *Journal of Researches* of 1839, renamed *The Voyage of the Beagle,* was published on his return and made him something of a celebrity among naturalist circles. Wallace was thrilled at the prospect of being useful to a scientific figure of rising prominence. More to the point, Darwin was one of the few connected naturalists Wallace was in communication with. So it was to Darwin that Wallace turned in June 1858 when he wanted feedback about his theory of evolution. Wallace was less concerned with Darwin's opinion on the matter than with that of Darwin's close friend, the eminent geologist Charles Lyell (1797–1875). Wallace asked Darwin to forward his letter to Lyell if he felt the paper worthy.

It is certain that had it not been for Wallace's letter, which according to Darwin himself 'could not have made a better short abstract' of his own theory, the *Origin* would not have been published in the form it took, nor would it have appeared in 1859.[9] Darwin was working on a much larger study of evolution – his 'big species book' as he called it – one for specialists that would have been multiple volumes. He believed that the only way to make the case for evolution was through an immense marshalling of facts presented at the highest possible level, 'a kind of science by filibuster that would overwhelm his opponents by fact and footnote.'[10] The arrival of Wallace's letter made publishing an abstract of this (as yet) unpublished larger work necessary in order to gain priority on a theory that Darwin could see slipping from his grasp.

Darwin was no revolutionary. While he was seeking priority for the scientific work he had engaged in for more than two decades, he was not looking to smash received wisdom or to overturn the central tenets of Christian thought. Indeed, this is one of the reasons why he had not publicized his discovery of the theory of natural selection in 1838. He had been schooled in natural history at a time when the study of nature was closely and intricately interwoven with religious metaphors and connotations. While the 'hard' sciences such as physics and astronomy had largely shed their religious baggage, at least in scientific explanations, in the English-speaking world natural history had developed strong links with Anglican theology. For someone like the Cambridge professor of mineralogy and historian of science William Whewell (1794–1866), flora and fauna could not be studied without reference to a Divine Creator, given the necessary connection between the study of the earth and that of human existence.

Stretching back to the early modern science of Isaac Newton and Robert Boyle, a tradition of naturalist writing developed in Britain that studied nature in order to uncover the work of God's hands – an enterprise that necessarily kept the 'history' in 'natural history.' William Paley's work provided wonderful portrayals of nature, highlighting the perfection of every aspect of the earth's plants and animals, which led to syllogistic arguments about the evidence of design leading to evidence of a designer. Paley's *Natural Theology* became the orthodox framework within which naturalists wrote, culminating in the Bridgewater Treatises of the 1830s, a series of eight individually authored scientific works on 'the Power, Wisdom, and Goodness of God as Manifested in the Creation.'[11] While these volumes signified a high-water mark for the tradition of natural theology, it is also clear that that tradition itself was not as dominant nor as homogenous as has previously been assumed.[12] Even so, most if not all natural theologians were theists – that is, they assumed not only that God had created the heavens and the earth and all its creatures, but also that God was thoroughly engaged in His creations, manipulating and interfering as needed. While the theory of evolution tended to undermine this view of God's role in the natural world, Darwin sought to accommodate the religious sensibilities of his readers by offering a theory of evolution that at the very least allowed for the existence of a Divine Creator. Indeed, the *Origin*, far from being the secular text it is often presented as, establishes a theory of evolution from within a Christian framework.

Darwin was very careful to at least appear to be writing from within the tradition of natural theology. One need look no further than the first page where Whewell is quoted: 'But with regard to the material world, we can at least go so far as this – we can perceive that events are brought about not by insulated interpositions of Divine power, exerted in each particular case, but the establishment of general laws.' Darwin then reaches far back and quotes Francis Bacon: 'To conclude, therefore, let no man out of a weak conceit of sobriety, or an ill-applied moderation, think or maintain, that a man can search too far or be too well studied in the book of God's word, or in the book of God's works; divinity or philosophy; but rather let men endeavour an endless progress or proficience in both.'[13] By the second edition, Darwin had chosen a third quotation, this one from the eighteenth-century bishop and theologian Joseph Butler: 'The only distinct meaning of the word "natural" is *stated, fixed* or *settled;* since what is natural as much requires and presupposes an intelligent agent to render it so, *i.e.*, to effect it continually or at stated times, as what is supernatural or miraculous does to effect it for once.'[14]

Darwin had carefully chosen these quotations. All three writers were symbolic of the connection between Christianity and modern science; thus, Darwin was subtly suggesting that the *Origin* was not inconsistent with Christian science. Furthermore, each quotation tends to support the image of God that appears throughout the *Origin* – that of a deity who governs the world through fixed laws rather than through constant supernatural interventions. There is no better indicator of this than the final sentence of the *Origin:* 'There is grandeur in this view of life, with its several powers, having been originally breathed into a few forms or into one; and that, whilst this planet has gone cycling on according to the fixed law of gravity, from so simple a beginning endless forms most beautiful and most wonderful have been, and are being evolved.'[15] Not only is Darwin here suggesting that the law of evolution is analogous to the law of gravity, a long-established scientific fact, but he is also suggesting that it was God who originally 'breathed' life, and by extension evolution, 'into a few forms or into one.' When the second edition appeared in December of 1859, Darwin – already on the defensive from both natural historians and natural theologians – was even more explicit in linking an evolving world with a divine being, changing the final sentence to read that 'there is grandeur in this view of life, with its several powers, having been originally breathed *by the Creator* into a few forms or into one.'[16]

When Darwin was a young man, naturalists tended to be clergymen first and men of science second. Indeed, if one was not independently wealthy, the only way to support a career of science was by attaining an academic post, and there were few outside of Anglican Oxbridge. While Darwin certainly had the means to support himself (his mother was a Wedgwood), his father, a medical doctor, was adamant that Charles become a professional. This resulted in failed attempts at law and medicine, which left a clergyman's life as the only option. He attended Cambridge from 1828 to 1831 with the purpose of becoming a Church of England cleric. While at Cambridge he was heavily influenced by Reverend Professor John Stevens Henslow (1796–1861), a leading botanist and geologist. Henslow prescribed for Darwin a reading list that included travel literature and books on natural science as well as works of theology. Darwin remembered in particular devouring the works of Paley, and he believed he 'could have written out the whole of the [Paley's] *Evidences* [1794] with perfect correctness.'[17]

While Darwin's early training in natural science was from the orthodox school of natural theology, his family had a history of scientific heterodoxy. His grandfather, the poet and physician Erasmus Darwin (1731–1802), was one of the first to popularize the idea of transmutation, and his *Zoonomia* (1794–96) was a beautiful but frightening poem about the possibilities of change inherent in all living things. He was also a member of the Lunar Society of Birmingham and believed in terrifying scientific practices such as the use of electricity – a new discovery at the time – a power that he believed could be harnessed to reanimate the dead.[18] Mary Shelley would later rely on Erasmus's authority to further convince her readers about the non-fictional merits of Victor Frankenstein's creation. 'The event on which this fiction is founded has been supposed,' Percy Shelley claimed in the preface to his wife's *Frankenstein* (1818), 'by Dr. Darwin ... as not of impossible occurrence.'[19]

The Darwins, then, had a family history connected with weird science, and if Charles's inquiry into evolution indicates a strong-willed devotion to the uncovering of truth, his continued belief in the water cure is a sign of the more tenuous and perhaps failed pseudo-sciences the Darwins also had a history of promoting. While the scientific and religious elites were debating his theory in Oxford, Darwin was indulging in his other scientific belief, letting the cold water ease the pain and anguish likely caused by his fears of how people would react to his transmutationist theories.

Darwin expected a backlash against his book, and he was prepared for one. The battle at Oxford in 1860 was only the most public of the debates over Darwin's book. Reviews, both negative and positive, had by this time flooded the periodical press, and Darwin was prepared for most of them, but one in particular caught him off guard: Richard Owen (1804–1892), who was Britain's leading comparative anatomist and Hunterian Professor at the Royal College of Surgeons, wrote a scathing anonymous review in the April 1860 issue of the *Edinburgh Review*. While Darwin and Owen had their differences, they were on friendly terms; indeed, the two of them had discussed the *Origin* at length prior to the review's publication, and Owen had been much more sympathetic to Darwin in person than in his review. Owen dismissed the idea that natural selection could do what Darwin claimed and suggested alternative possibilities, such as his own theory of archetypes, a theory that postulated a series of Platonic forms acting as a general blueprint for the earth's many species. This theory of species development was largely accepted in natural-theological circles, and Owen made it sound as if all of the evidence that Darwin had provided for his evolutionary theory was simply further evidence in favour of archetypes. Owen went on to ridicule Darwin's writing style and his friends (Hooker and Huxley); he even went so far as to question Darwin's competence.[20] It was difficult for Darwin not to take all of this personally.

Darwin did not find much use for Owen's attack except for Owen's implication that he had not properly acknowledged indebtedness to his evolutionary predecessors. Owen, of course, was suggesting that

Darwin had stolen many of his ideas in coming to the wrong-headed theory of natural selection. Whatever Owen's motives for making the criticism, Darwin believed that this was a constructive criticism and worth acting on.

Darwin had earlier been privately rebuked by Baden Powell (1827–1860), Savilian Professor of Mathematics at Oxford (and father of the Boy Scouts founder), for not properly acknowledging his predecessors. Here, Powell was largely referring to his own work on evolution. At the time, however, Darwin argued that writing a proper historical analysis of evolution was not necessary. He had 'already acknowledged with pleasure all the chief facts and generalisations which I have borrowed,' and besides that he believed that properly writing about the authors of evolution would be a taxing process best left 'to the Historian of Science rather than to me.'[21] Later that day, however, Darwin would again write to Powell, this time about a preface he had already written that was for his 'big species book,' a preface he had completely forgotten about. From it, he quoted a passage in which he listed Powell's *Essays on the Unity of Worlds* of 1855 as treating the 'Philosophy of Creation ... in an admirable manner.'[22] Perhaps writing a historical sketch would not be so difficult after all.[23]

There was certainly precedent for doing such a thing. Charles Lyell in his *Principles of Geology* (1830) had included no fewer than four chapters on the historical development of geology; they served as a means to foreground his own work as the only truly scientific approach to studying geology. Lyell's primary target in his own 'Historical Sketch' – the title he gave this set of historical chapters for the third edition (1834) – was the theory of catastrophism – a favourite theory among natural theologians, one which posited that the earth's changes were the results of a series of large catastrophes in the past. The French zoologist Georges Cuvier (1769–1832), who was this theory's primary author, was largely silent on the theological aspects of catastrophism, but he did suggest that the last catastrophe on record was likely that of the Flood recorded in the book of Genesis. The geologist William Buckland (1774–1856) was Britain's foremost proponent of the theory, and he was much more outspoken about its theological dimensions. Lyell argued that advances in geology were being retarded by the likes of catastrophists such as Buckland, who speculated beyond what they observed, relying instead on the mythology of supernatural events. Lyell argued in contrast that changes in the earth's surface could be understood only by examining causes operating in the present – that is, by considering causes that can actually be seen. For such a theory to work, however, Lyell had to assume that the earth in the past was much like it is today, that there is a perpetual cycle of growth and decay, an extremely slow process of inorganic change. He was also adamant that there was no progressive development inherent in such geological change, in contrast to the teleological direction of history suggested by catastrophism.

Even though Lyell's theory, dubbed uniformitarianism, explicitly challenged the catastrophist view, he was deeply respected in scientific circles, for he was from a well-respected family and had many ties to the Cambridge elite. He was also a devout Unitarian. While he no doubt had his critics, Anglican heavyweights like Whewell and the Woodwardian Professor of Geology at Cambridge, Adam Sedgwick (1785–1873), were outspoken in their praise of *Principles* even though they were fundamentally opposed to the theory of uniformitarianism. The young Darwin had been profoundly moved by Lyell's work. He had read the first volume during his *Beagle* voyage and had had the second one sent to him on publication while he was still at sea. Thus he began to see the world through Lyell's eyes. At least two aspects of Lyell's work were fundamental to Darwin's understanding of organic species: the unlimited draughts of time posited by uniformitarian geology; and the immense change made possible by the multiplication of slow, relatively small forces.

Indeed, Darwin was quick to consider Lyell's work in relation to the earth's organic species, and he believed that his theory of evolution owed more to Lyell than to anyone else. However, Darwin's theory of natural selection owed something to the catastrophists as well. However slowly natural selection works, it is difficult to ignore its progressive nature in Darwin's formulation. 'As natural selection works solely by and for the good of each being,' Darwin writes in the *Origin*'s penultimate paragraph, 'all corporeal and mental endowments will tend to progress towards perfection.'[24] Lyell would eventually embrace Darwinian evolution, but only with regard to the earth's non-human species. Despite their differences, Lyell was firmly in Darwin's corner from the beginning. It was he who convinced Darwin to put his theory to paper, and later he organized a reading of Darwin's work alongside that of Wallace at the Linnaean Society on 1 July 1858 to ensure that Darwin maintained priority over natural selection. Lyell was an extremely important scientific figure for Darwin to have on his side, being a well-respected scientist with both the older and younger generations, with natural theologians as well as scientific naturalists. It is also clear that Lyell's 'Historical Sketch' influenced Darwin to write one of his own.

In 1856, when Lyell suggested to Darwin that he publish his work on species, he also suggested that Darwin do as he had done: write a history of the subject itself. His response was similar to the one he made to Powell in 1860: he felt that he 'could not attempt a history of the subject' but perhaps could write a few pages on the 'two or three leading and opposed authorities.'[25] The published version of Darwin's 'Historical Sketch,' subtitled 'Of the Progress of Opinion on the Origin of Species, Previously to the Publication of the First Edition of this Work,' was much longer and more detailed than he initially planned; even so, it was only eight pages compared to Lyell's forty-seven. It is closer to a literature review than a history of an idea, yet there is much to glean from these short eight pages, and not just about the history of pre-Darwinian evolutionary theory or about Darwin's own understanding of, and place within, that history. More important, we learn how Darwin wanted his theory and himself to be remembered.

Darwin begins his 'brief sketch' by suggesting that 'until recently the great majority of naturalists believed that species were immutable productions, and had been separately created. This view has been ably maintained by many authors.' Clearly, he is avoiding Lyell's polemical approach. But he then admits that he was not the first to propose a transmutationist theory; a few naturalists over the years 'believed that species undergo modification and that the existing forms of life are the descendents by true generation of pre-existing forms.' Before discussing these authors, Darwin justifies excluding those who have made mere 'allusions to the subject,' such as classical authors. Thus he relegates Aristotle to a footnote, even though he has already admitted in his work that we can 'see the principle of natural selection shadowed forth'; and he dismisses the French naturalist Buffon (1707–1788) for not entering 'on the causes or means of the transmutation of species' despite being the 'first author who in modern times has treated it in a scientific spirit.' Darwin also relegates the evolutionary views of his grandfather, Erasmus, to a footnote, continuing his long practice of minimizing any familial influences.[26]

Darwin's historical sketch really begins with the French naturalist, Jean-Baptiste Lamarck (1744–1829), 'the first man whose conclusions on the subject excited much attention.' As explained by Darwin, Lamarck's 'celebrated' work 'upholds that all species, including man, are descended from other species.' Lamarck was led to this view by his observations of a 'gradual change in species, by the difficulty in distinguishing species and varieties, by the almost perfect gradation of forms in certain groups, and by the analogy of domestic productions.' The key mechanism of change, for Lamarck, was the notion of 'acquired characteristics,' a mechanism built into species that allowed them to change on the basis of use and disuse 'such as the long neck of the giraffe for browsing on the branches of trees.'[27]

Darwin was very careful to distance his own theory from Lamarck's, which had a bad name throughout Britain's scientific and religious establishment. Darwin emphasized chance and contingency in his theory of evolution, and he downplayed the possibility of acquiring new characteristics. He also avoided Lamarck's theory of spontaneous generation, which suggested that new species could simply come into being out of thin air. Evolution, argued Darwin, required many generations for the most minor of changes to take place; here he was rejecting the abrupt generational change posited by Lamarck. The chief mechanism of change for Darwin was natural selection – that is, the struggle between individuals of the same species whereby those individuals best adapted to their environment survived, leaving more offspring and thereby perpetuating their own characteristics in succeeding generations. Under the terms of Darwinian evolution, new species would only appear after this very slow and gradual accumulation of advantageous variations over many generations. Natural selection involved a constant 'struggle for existence' in which death and extinction were fundamental facts of life – indeed, central to the process. Lamarck, it should be noted, did not believe in extinction.

At this point in Darwin's historical sketch it is clear that he is trying to set up a fairly long history of evolution while establishing the originality of his causal mechanism, that being natural selection. Ironically, once physicists like Lord Kelvin began later in the century to suggest that the earth was not nearly as old as Darwin needed for natural selection to do its work, Darwin brought Lamarck back from the grave; in later editions of the *Origin* he accorded much more power to the process of acquired characteristics in order to hurry evolution along. In 1860, however, Darwin wanted to put as much distance between natural selection and Lamarck as possible, given that just about anything from post-revolutionary France was seen by the British as radical and suspect. It was bad enough that evolution was already associated with working-class radicalism, materialism, and atheism.[28]

Next on the list of Darwin's evolutionary precursors was Étienne Geoffroy Saint-Hilaire (1772–1844), another French naturalist, who expanded and defended Lamarck's theory of evolution but was much more 'cautious in drawing conclusions' – a trait Darwin no doubt appreciated. Unlike Lamarck, however, Geoffroy Saint-Hilaire believed that evolution was the result of 'the conditions of life, or the *monde ambient*' rather than anything internal; he also 'did not believe that existing species are now undergoing modification.'[29]

Darwin continues by listing some evolutionists who were less well known in his day and, even more so, in ours. A Dr W.C. Wells delivered a paper at the Royal Society in 1813 that described the evolution of certain aspects of human traits; he also seemed to have in mind a mechanism of change quite similar to natural selection. The Honourable and Reverend W. Herbert argued that individual species 'were created in an originally highly plastic condition' and that by intercrossing and by variations they have produced 'all our existing species.' Patrick Matthew (1790–1874) is given credit for providing an almost exact summary of natural selection in 'scattered passages in an Appendix' to a work on 'Naval Timber and Arboriculture' (1831). Darwin is unable to 'understand some passages,' but he admits that Matthew 'clearly saw ... the full force of the principle of natural selection.'[30]

The Scottish zoologist Robert Edmund Grant (1793–1874) is given credit for his 'well-known' 1826 paper in the *Edinburgh Philosophical Review* in which he declares, in the final paragraph, 'that species are descended from other species, and that they become improved in the course of modification.' Darwin did not learn of Grant's evolutionary views until after the first edition of the *Origin* was published, as he had of Matthew's, Herbert's, and Wells's. Indeed, Grant was one of Darwin's professors during his second year of medical studies at Edinburgh University in 1826. The paper to which Darwin refers was published while he was a student of Grant's, and there is no doubt that Darwin was introduced to Lamarckian evolution through Grant's teaching.[31]

Yet the most popular British evolutionist prior to the publication of the *Origin* was not Grant or any other British evolutionist Darwin mentioned but rather the anonymous author (Robert Chambers [1802–1871]) of the immensely popular *Vestiges of the Natural History of Creation* (1844). The subject matter for Chambers was not just the evolution of species; it was also the formation of the solar system, the creation of life on earth, and even the development of the human mind. His was a form of Lamarckian evolution applied to the history of life as we know it.[32] The book, it is safe to say, caused a 'sensation,' going through ten editions in ten years and fifteen throughout the nineteenth century. By 1860, 23,700 copies of *Vestiges* had been published. By 1900 the figure would rise to 40,000. These were huge numbers for a work of science; when it was first published it took hold of the popular imagination thanks to Chambers's populist writing style as well as his romantic narrative, which posited a steady, progressive development of humanity that would eventually reach perfection.[33]

While the general reading public found much to admire in *Vestiges*, the sensation wrought by the book caused the upper classes as well as the scientific and religious establishment to lash out. Many of the figures who later would clash over the *Origin* were united in their opposition to *Vestiges*. Thomas Henry Huxley, the protagonist of the Oxford debate, wrote many damaging critiques of *Vestiges* in which he challenged not just the scientific merits of the work, but the general readership that *Vestiges* sought out – a readership that could not possibly tell good science from bad. For Huxley, *Vestiges* was the work of an amateur masquerading as a specialist; furthermore, the anonymous author was doing much harm to the public's understanding of scientific work.[34]

Darwin had his own misgivings about *Vestiges* – in particular, it lacked observable data to support its general theory, and it promoted a Lamarckian brand of evolution. More to the point, the reaction of many of Darwin's close friends and colleagues – particularly Lyell – suggested to him that Britain was not quite ready to consider an evolutionary perspective, no matter how scientific. He had written a short manuscript summarizing his evolutionary findings, and now he felt compelled to shelve it, telling his wife Emma to publish its contents only upon his death. His fear was that he would be caught up in the furor that *Vestiges* had unleashed.

At the BAAS meeting in Cambridge in 1845, *Vestiges* was repeatedly denounced. The association's president, the astronomer John Herschel, began the onslaught during his presidential address. Works such as *Vestiges* needed to be denounced upon arrival, he argued, because they threatened the 'brotherhood' of science, a brotherhood united in the study of 'the wonderful works of God.'[35] Throughout the week-long proceedings, other members of the brotherhood attacked *Vestiges*. The assailants included the Bishop of Oxford, Samuel Wilberforce, who denounced the work as an attempt to link irreligion with science, a link that surely no one would attempt to prove at Cambridge, of all places.[36] In retrospect, the attacks unleashed on *Vestiges* had, in the words of historian Loren Eiseley, 'a quite unreal character'; criticisms of the theory were launched at multiple levels, from its very details to its metaphysical and teleological implications.[37] Darwin surmised, perhaps correctly, that this was not the time or the place to unveil his own evolutionary theory. He waited sixteen years, and even then it was Wallace's letter that forced him to act. By then he had convinced many of his close friends of his theory's validity, thereby gaining much-needed support for the backlash that was sure to follow.

In his historical sketch, Darwin makes a point of criticizing *Vestiges*. By then, many negative reviews have lumped Darwin's work alongside that of Chambers, which at the time is still a popular book, and Darwin here is taking the opportunity to further distance himself from it. *Vestiges*, Darwin argues in his sketch, suggests that evolution occurs by 'sudden leaps, but that the effects produced by the conditions of life are gradual.' Darwin writes that he cannot see how these 'two "impulses" account in a scientific sense for the numerous and beautiful co-adaptations which we see throughout nature.' The recently published tenth edition of *Vestiges*, he goes on, is a great improvement over previous editions, which displayed 'little accurate knowledge and a great want of scientific caution.' And Darwin is compelled to admit that the work is written in a 'powerful and brilliant style' that has contributed to its 'very wide circulation.' Indeed, despite *Vestiges*' many faults, it has done an 'excellent service in this country in calling attention to the subject [of evolution], in removing prejudice, and in thus preparing the ground for the reception of analogous views.' Darwin may have owed *Vestiges* very little when it came to the finer scientific points of evolution, but he did owe *Vestiges* a great deal when it came to preparing not just the public but also the scientific and religious communities for his own, far more cautious work on the subject.[38]

Owen in his review suggested that Darwin's contribution to evolutionary theory was minimal. He also intimated that Darwin had not properly acknowledged his debt to other evolutionists. And he suggested that Darwin had added very little that was new to the question, 'leaving the determination of the origin of species very nearly where the author found it.' Owen's subtle point was that Darwin had wrapped old ideas in the guise of originality.[39]

Darwin, much like Chambers before him, relied on many of Owen's universally well-regarded anatomical observations in building his case for evolution; yet he also lumped Owen in with the immutability-of-species camp while not giving him any credit for his rival theory of evolution. Not that Owen was an outspoken evolutionist. His theory of archetypes, while embraced by many anti-evolutionary Creationists, was a theory of evolution as far as Owen was concerned. Much of his review was a summary of his own archetype theory in contrast to natural selection, which was a process that could not accomplish what Darwin claimed. In Owen's view, evolution was the product not of chance variations but of a divine plan whereby sudden mutations occur that are consistent with an original archetype. Throughout the review, Owen portrayed himself, in the words of historian Nicolaas Rupke, as a 'misrepresented and ignored evolutionist' – that is, as misrepresented and ignored by Darwin in the *Origin* and then eclipsed by him in popular opinion.[40]

Darwin used his historical sketch to respond to Owen's review while adding Owen to his list of evolutionary precursors. In it, Darwin says that he was deceived by Owen's seemingly anti-evolutionary expressions, such as 'the continuous operation of creative power,' and that he wrongly 'included Professor Owen with other paleontologists as being firmly convinced of the immutability of species; but it appears that this was on my part a preposterous error.' While Darwin seems to be admitting an error, he is actually suggesting that Owen is confusing on the issue. Later, Darwin argues that his 'initial inference still seems ... perfectly just,' given other similar passages of Owen's. Darwin also claims that extracts of the correspondence between Owen and the editor of the *London Review* have confused him further, because in those, Owen 'claimed to have promulgated the theory of natural selection before I had done so.' Darwin admits that he was surprised and yet satisfied 'at this announcement' only to realize on the basis of Owen's recently published work that he 'had either partially or wholly again fallen into error.' Anyone familiar with Darwin's published writing knows that this is as vicious and as underhanded as he ever gets. Despite admitting to misunderstanding Owen, he is making the case that it is Owen who is confusing on the matter, that his writings are contradictory, and that perhaps jealousy rather than science has guided Owen's posturing, which is both confusing and defensive. 'It is consolatory to me,' concludes Darwin, 'that others find Professor Owen's controversial writings as difficult to understand and to reconcile with each other, as I do.'[41]

Darwin goes on to very briefly describe the evolutionary views of the French naturalist Isidore Geoffroy Saint-Hilaire (the son of Étienne), as well as those of Dr Henry Freke; the future social Darwinist Herbert Spencer; the French botanist Charles Naudin; the 'celebrated geologist' Count Alexandre Keyserling; the Professor of Anatomy at the University of Bonn Hermann Schaaffhausen; the French botanist Henri Locoq; Baden Powell; Alfred Wallace; and Karl Ernst von Baer, 'towards whom all zoologists feel so profound a respect.'[42] This is by no means an exhaustive list of evolutionists or of those who had made evolutionary statements at one point or another. However, the reader cannot help but be taken aback by the many evolutionary precursors Darwin acknowledges; and even as he begins with Lamarck, he admits that there were likely many others stretching back to the ancient Greeks. It is clear, however, that throughout his sketch Darwin is concerned with establishing priority over the theory of natural selection. While a few authors clearly anticipated the theory, theirs tended to be scattered thoughts in the context of poorly worked-out evolutionary theories. Wallace was the only one who stated a similar theory with adequate evidence, but Darwin makes it clear that this was much later than his own discovery of the theory of natural selection.

Darwin ends his sketch by mentioning the work of his two great friends and colleagues, Thomas Henry Huxley, Professor of Natural History at the Royal College of Mines; and Joseph Hooker, the young botanist and assistant director of Kew Gardens. During the late 1850s, Hooker and Huxley were members of Darwin's circle of close friends, and they were let in on Darwin's evolutionary secret, eventually becoming supporters, defenders, and contributors to the theory's development. After Darwin's initial paper was read to the Linnaean Society, and prior to the publication of the *Origin*, both Hooker and Huxley published works supporting the theory of evolution. It is fitting, then, that Darwin ends his sketch with an acknowledgment of the work that he had influenced even before the *Origin* hit the bookshops.[43]

When the *Origin* was finally published, he could count on his friends to defend his theory, often in the context of their scientific specialties. Huxley was particularly outspoken in his support, writing no less than three reviews of the *Origin*, in the *Westminster Review* (1860), *Macmillan's Magazine* (1859), and, most importantly, in the *Times* (1859). Hooker, for his part, wrote an extremely favourable review of the

Origin in the *Gardeners' Chronicle* (1860). In Darwin's absence, Huxley and Hooker would prove extremely valuable to the cause of Darwinian evolution. He would indeed need their assistance.

While Darwin had defenders, he also had outspoken critics. Next to Owen and Wilberforce, Darwin's fiercest critic was likely the geologist Adam Sedgwick. Darwin had been one of Sedgwick's students, and the two remained friends despite their scientific differences. Darwin had the publisher send Sedgwick a copy of the *Origin,* and Sedgwick thanked Darwin despite having 'read your book with more pain than pleasure.' Sedgwick was of course a seasoned geological controversialist, having sparred with Lyell – among others – on the evidence for uniformitarianism, a geological theory that underpinned much of Darwin's work. Much as he had with his critique of Lyell, Sedgwick admitted to admiring parts of the book, before adding that other parts made him laugh 'till my sides were almost sore' and that he had read much of the book with 'profound sorrow.' He believed that Darwin had *'deserted'* the 'true method of induction' in that many of his conclusions could not be proved or disproved. Sedgwick believed that Darwin's 'grand principle' of natural selection was nothing more than a cause of secondary consequence. It would be better to think of 'development' rather than 'evolution,' he argued, and to view any causation as the act of God: 'I can prove that He acts for the good of His creatures. He also acts by laws which we can study and comprehend – Acting by law, and under what is called final cause, comprehends, I think, your whole principle.' Sedgwick concluded his letter by referring to himself as 'a son of a monkey and an old friend' and that the letter had been written in the spirit of 'brotherly love.'[44] Despite the friendly tone, Darwin felt that Sedgwick's criticisms were 'childish' and that they indicated he likely would be 'greatly abused' in future critiques.[45] Indeed he was.

Of course not all of the debate was relegated to personal letters and the periodical press. The meeting of the Cambridge Philosophical Society (CPS) in May of 1860 foreshadowed the reception of the *Origin* at the BAAS meeting the following month. At the CPS meeting, Sedgwick made his criticisms public by savaging the *Origin* as a heretical text, grouping it with the recently published *Essays and Reviews,* a collection of essays of biblical criticism written by liberal Anglicans that challenged strict Anglican doctrine (see chapter 2). Darwin wanted to avoid precisely the analysis that Sedgwick produced: that the *Origin* was part of a broad campaign to appropriate science for the purpose of undermining the Church of England. Darwin's former mentor Henslow happened to be present and defended his former student's right to study the species question. 'I got up,' Henslow wrote to Hooker, 'and stuck up for Darwin as well as I could, refusing to allow that he was guided by any but truthful motives.' Henslow went on to explain, however, that 'he himself believed that [Darwin] was exalting and not debasing our views of the Creator, in attributing to him a power of imposing laws on the Organic World by which to do his work, as effectually as his laws imposed upon the inorganic had done it in the Mineral Kingdom.'

Henslow asked Hooker, once he finished reading the letter, to send it to Darwin, knowing that Darwin wished 'to know of all criticisms, pro and con.' Henslow then listed a series of recent reviews, most of them negative, which Darwin should read.[46] Darwin did not want to bury his head and hide from the criticism; he was more than happy to consider criticism of the constructive variety in order to improve future editions of the *Origin*. However, much if not most of the criticism was dismissive rather than helpful, as many reviewers had rejected his theory out of hand for religious reasons. Henslow's strategy in defending Darwin against such critique was to highlight the theological cues in the *Origin* in order to show that Darwinian evolution need not be inconsistent with natural theology. The following month, Huxley would offer quite a different approach at the BAAS meeting in Oxford.

In light of the fact that the debate over Darwinian evolution was expanding from the periodical press onto the much more volatile terrain of public meetings, Darwin felt that his interests would best be served by avoiding such meetings, at least in the immediate aftermath of the *Origin*'s publication. Darwin claimed to suffer from a variety of ailments that kept him from leaving his house in the village of Downe for more than a few days. He complained of constant stomach pains and flatulence, and he often had to excuse himself to other rooms when visiting friends, sometimes for hours on end, to attempt to relieve his stomach cramps. He also suffered bouts of dizziness and general weakness, staying in bed for days at a time. He believed himself to be an invalid; his health was constantly at the forefront of his mind.

He had discovered hydrotherapy in the 1840s and would often travel quite far to receive his water cure. It helped that several days of rest and relaxation was part of these treatments. Darwin used these excursions to get away from his day-to-day struggles. Whether or not hydrotherapy was the cure Darwin claimed it to be, it is clear that these trips did much to relieve the stress he felt. It is not surprising that he chose the week of the BAAS meeting in June to receive treatment for his illness.

Darwin had already defended his theory through his historical sketch, which was written about a month prior to the Oxford debate. The sketch was a subtle polemic against Owen; it also allowed Darwin to distance his own theory of evolution from the rival theories offered by Chambers and Lamarck. In this way he acted as historian, presenting his own theory of natural selection as original (notwithstanding his precursors) and as a cautious and fully developed treatment of the topic at the end of a long line of evolutionary theories and passages, a history of which his book was the culmination.

He had planned to travel to Oxford with Hooker, even making inquiries into private lodgings.[47] However, his earliest correspondence about the upcoming BAAS meeting shows him unwilling to commit himself. He wrote to both Lyell and Hooker describing a willingness to attend, but several weeks before the meeting he was already setting the stage for his absence. His daughter Etty was not well and Darwin could envision health problems of his own. 'I do not know what to say about Oxford,' he wrote to Hooker on 15 May, a full six weeks before the BAAS meeting. 'I should like it *much* with you; but it must depend on health.'[48] By 30 May, Darwin was 'in great doubt about Oxford.' Etty was much improved, 'my doubt being chiefly from my own health.'[49] On 6 June, he told Lyell he might see him in Oxford.[50] But on 12 June, he wrote to Hooker that the Oxford trip was 'extremely doubtful,'[51] and on 26 June he finally backed out: 'My stomach has utterly failed; and I cannot think of Oxford; on Thursday I go for [a] week of water-cure to "Dr Lanes, Sudbrook Park, Richmond Surrey."'[52] While the debates about his work were reaching a fever pitch, his water cure was calling. Darwin's friends would have to go it alone.

2 The Struggles of Soapy Sam

Give up to God the time before given to her.

Samuel Wilberforce, 19 March 1841
(on the death of his wife)

Soapy, ***a***
5. *slang.* **a.** Ingratiating, suave, unctuous.
1854 E. TWISLETON *Let.* 22 June (1928) xi. 202 The Bishop of Oxford I never do like ... his manner, when Lords are in presence, richly merits his popular sobriquet of 'Soapy Sam'. **1865** *Pall Mall G.* 28 Oct. 5 But why ... do people call him [Bp. Wilberforce] Soapy Sam? **1910** *Blackw. Mag.* Feb. 182/2 He had once been famous for his soapy manners.

Oxford English Dictionary

Subtlety was not a characteristic generally associated with Samuel Wilberforce. Like his father, William (1759–1833), 'the Great Emancipator,' who struggled against the slave trade as leader of the evangelical Clapham Sect, Samuel fought tooth-and-nail to defend his beliefs. He was greatly influenced by his father's deep devotion to the evangelical faith, and he showed this by constantly searching his soul to find moral codes worth fighting for. Unlike his father, however, the struggles of Soapy Sam appear to have been on the wrong side of history. Outside of the ecclesiastical history of Victorian England, Samuel is known only as a minor character in the struggle between science and religion, in which he played the role of villainous defender of faith against the rational proponents of evolutionary theory. Even his nickname has suffered from posterity, at one time suggesting that he could escape the dirtiest of debates without a trace of mud on his person. Unfortunately for Samuel, he did not escape the Oxford debate clean, as it were. It was no longer his person that was somehow cleansed by his soapy nature; it was his manners that were now seen as saponaceous.

But we must be fair to Samuel Wilberforce. While he was no doubt, as Bernard Lightman put it, 'a pale evangelical imitation of his father,'[1] it is important to understand that as far as Samuel himself was concerned, the same evangelical guidance that led William to heroically struggle against the slave trade and then slavery itself was what motivated Samuel in his less than heroic struggle against evolution. As well, the death of his wife in 1841 engendered a crisis of faith that Samuel overcame through a renewed devotion to the Church of England, that led him to embrace still more fervently the Anglican faith as his life work. His battle against the Darwinian theory of evolution must be understood in the context of a broader struggle within the Church of England engendered by the rise of biblical criticism. Evolution implicitly undermined literal readings of the Bible – in particular, that of the history of the earth found in Genesis – but remember here that biblical scholars relying on new methods of historical textual analysis were by then explicitly challenging the words of the Bible by reading them against what historians, geologists, and archaeologists were saying actually happened. And for the most part, the stories told in the Bible, it was found, either simply did not happen, such as the Flood, or did not happen in the way described, such as the life of Jesus Christ. None of this would have been a problem for Wilberforce except for the fact that many clergymen were beginning to accept many of the findings of biblical criticism. Worse still, they were seeking to change the central tenets of Christianity to accommodate the findings of that scholarship. From Wilberforce's perspective as Bishop of Oxford, this was heresy.

While Wilberforce became the face of an orthodox Christianity against the rising tide of the findings of a rational science, his role could not have been forecast. Indeed, for most of his career as a clergyman he had mediated between opposing factions of the Church, largely espousing the most practical solutions to the problems created by the infighting. The evangelical revival of the late eighteenth century, of which Wilberforce's father was both a product and a symbol, heavily influenced Samuel's belief system. The evangelical movement was a response to the rise of Enlightenment thinking, which called for people to rely entirely on the rational mind and to reject faith-based knowledge. In response, evangelicals preached an individual relationship to God, emphasizing daily prayer and personal conversations with Him. Scripture was understood as the actual words of God and for the most part was read literally. It was an evangelical's duty to conform to God's will as found in both scripture and conversation. In this way an evangelical prepared for his or her release from this world in order to be saved in the next. Eternal salvation was the guiding force behind Wilberforce's evangelicalism; and despite the challenges facing the evangelical faith – challenges that caused two of Samuel's brothers to reject their father's religious teachings – Samuel himself never abandoned its core.[2]

That evangelical faith was preached to Samuel by his father throughout his youth. When Samuel turned twelve, William began writing him letters whose purpose was to provide moral guidance for the developing young man. There are more than six hundred of these letters, which Samuel carefully preserved, and they served as the foundation of his moralistic belief system. In a letter dated 23 March 1822, when Samuel was sixteen, William asked him to *remember* all that he had learned from his father. 'I am always tempted to conclude my letters with Charles I.'s last word REMEMBER,' he wrote to Samuel, 'which may naturally be supposed to refer to whatever the speaker is known to have desired to live in the recollection of the person addressed.' Samuel, of course, knew what his father meant by 'remember.' Remember a father's wishes for his son's 'temporal and eternal happiness, and endeavour to have them realized.' Remember to 'watch unto prayer' and to 'maintain such a state of mind' that will 'render you fit at any time … to compose your spirits and engage in that blessed exercise.' Remember to 'walk by faith and not by sight' and '"to do all in the name of the Lord Jesus," that is, to bear in mind that He is always present with you, that He witnesses all your thoughts, words, and actions, and that as His servant, His friend, His purchased possession, you ought always to be living to His glory.'[3] A.R. Ashwell, author of the first volume of the *Life of Samuel Wilberforce*, compared these letters from William to his son with those written by Samuel as an adult and found that William had clearly 'exercised the most powerful influence on the formation of [Samuel's] character … Samuel was indeed his father's son.'[4]

A few days before Samuel went off to study at Oxford, William wrote another letter in order to mentally prepare his son for his new journey. 'You are the son, my dearest Sam, of parents who I can truly declare have made your eternal interests the grand object of their care, and who on this principle selecting not only your tutor, but as far as possible your associates also, have endeavoured to preserve you pure from all contagious influences and from corrupt associates.' Now was the time for Samuel to become his 'own master,' and William wanted to impress on him the importance of his future independent choices, which would determine 'what is popularly called your character.' However, William continued, 'you, as *my* son, will be tried to a different standard from that which is commonly referred to, and be judged by a more rigorous rule; for it would be folly, rather than merely false delicacy, to deny that from various causes my character is more generally known than that of most men in my rank in life.' Samuel, in other words, would have to live up to an ideal image to reflect his father's reputation. 'Remember my boy,' William concluded, 'that you have my credit in your keeping as well as your own.'[5] This was certainly a lot to ask of a young man, who like his brothers had been sheltered from any sort of religious alternative throughout his young life. William believed he was preparing his sons for the vices and controversies of the outside world by indoctrinating them into the evangelical faith. By sending Samuel to Oxford in the 1820s, however, William was testing Samuel's devotion much more than he likely realized.

When Samuel Wilberforce attended Oriel College at Oxford University in the 1820s, several key members of what would become the Tractarian (or Oxford) Movement were fellows, most notably John Henry Newman. Tractarians believed that the revival of evangelicalism, with its emphasis on one's personal relationship with God and the Bible, had necessarily weakened the spiritual *and* corporate roles of the established Church. In the 1830s, through its series of publications titled *Tracts for the Times* (1833–41), the Tractarian Movement gained a massive following and became a serious challenge to evangelicalism. The movement sprang into action following the election of a Whig government headed by Lord Grey in 1830. One of the planks in the Whig platform was church reform, which thrust the relationship between Church and State onto the national agenda. The Tractarians abhorred this threat to the sovereign power of the Church. The Church of England had been the official national church since the sixteenth century. It enjoyed state funding but was free of direct government intervention – or so it seemed until Lord Grey began making his threats. In response, Tractarians sought to establish the Church of

England as a separate entity with an authority based not on the state apparatus but on the apostolic succession of the bishops.

Samuel was not unsympathetic to the Tractarian Movement. He never wholeheartedly adopted the label – unlike his brother Robert, who along with Newman later followed the logic of the movement by converting to Roman Catholicism. But at the same time, he was committed to saving souls, and if increasing the sovereignty of the church could help in that endeavour he was more than happy to support any policies that did so. In other words, Samuel was as pragmatic as possible when it came to dealing with the conflicts between the High (Tractarian) and Low (evangelical) Church, and he acted as something of a mediator between them, keeping one foot in either faction, supporting one side or the other depending on the circumstances. Samuel's primary goal was to promote the Anglican faith, and he did not care that this meant supporting the Tractarians one day and the evangelicals the next.

Where Wilberforce did draw the line, however, was in challenges to what he believed to be fundamental Christian truths found in the Bible. He was deeply sceptical of those looking to rationalize a religion based partly on faith, and he was outspoken in condemning such heresy. Wilberforce felt that the extreme views of the Tractarian Movement ran in just that direction. In an 1838 letter to his friend Charles Anderson, he stated that he agreed with many beliefs of the Tractarians, but he also admitted to 'some fears':

> My principal fears are, that they will lead to the depression of true individual spirituality of mind in the reaction of their minds from the *self-idolising tendency* of the late leading religious party, by leading others to elevate solely the *systematic* and communion parts of Xty.; ... I cannot use all their language about the Eucharist; I cannot bear Pusey's new sin after baptism. They hold up a glorious standard of holiness, and for *us*, my dear Charles who know well the hopes of the Gospel, and can supply all they leave deficient, it is the very thing needful; but there are ignorant and bowed-down souls who need a more welcoming treatment than their views of penitence will allow.[6]

Samuel clearly remembered his father's words 'to walk by faith and not by sight' – a phrase that would guide him through his later opposition to evolution. It is significant that the battle Wilberforce would fight against evolution was related to battles he was engaged in within the Anglican faith.

Wilberforce's fears of Tractarian extremes sucking the life out of a spiritual Christianity were somewhat fulfilled in the 1850s and 1860s as the liberal Broad Church Movement gained a following in apparent reaction to the rational extremes of the Tractarians. It was in opposition to the Broad Church Movement, as the leader of orthodox Anglicanism, that Wilberforce began to be seen as a fierce and outspoken debater. The Broad Church Movement, in an attempt to modernize the Church, was constantly attempting to reshape Christianity to conform to science. This had the effect of uniting the evangelicals and the Tractarians in opposition to it. While both the evangelicals and the Tractarians posed their own particular challenges to certain Anglican procedures and beliefs, the rationale underpinning both was deeply Christian. Neither the evangelical nor the Tractarian in Wilberforce could find any common ground with the Broad Church Movement. In 1860, three months before Wilberforce denounced evolution at the Oxford debate, the Broad Church Movement published its *Essays and Reviews*. That book excited a scandal within the Church of England far greater than did Darwin's *Origin of Species*.[7]

Published on 21 March 1860, the mildly titled *Essays and Reviews* was a collection of seven articles by seven different authors, six of whom were – highly notably – well-known Anglican clergymen. Each of the seven articles attempted to argue for Christianity's relevance by showing how its central beliefs held true despite the recent findings of science and biblical criticism, which appeared to challenge many fundamental Christian truths. The authors were implying that the truth of Christianity was only obscured by orthodox theology, which rested on external evidences of miracles and prophecies. Such a dishonest foundation for a religion would surely only lead to large-scale rejection in the face of common sense. Indeed, the authors felt that they needed to bring Christianity into the modern world not by denouncing current scholarship that posed particular problems for Christianity's medieval relics, but rather by using the advancement of science as an opportunity to rid Christianity of its untenable beliefs.[8]

Baden Powell argued that the evidences of Christianity must be judged entirely by reason and intellect. He explained that if the scientists' consensus were followed with regard to the uniformity of nature, then no testimonial evidence could possibly be provided in support of the supernatural. It followed that it was not miracles but the absolute uniformity of nature that supplied evidence for Christianity. Science, argued Powell, shows us 'that beyond the domain of physical causation and the possible conceptions of *intellect* and *knowledge*, there lies open the boundless region of spiritual things, which is the sole domain of *faith*.' Indeed, he continued, the 'more knowledge advances, the more it has been, and will be, acknowledged that Christianity, as a real religion, must be viewed apart from connexion with physical things.' It should be noted that Powell praised Darwin's *Origin* in passing as a further proof of 'the grand principle of the self-evolving power of nature' – a comment he added during the proofing stage of his article.[9] Powell recognized in the *Origin* not an argument for the non-existence of God but, rather, an argument in favour of an all-powerful uniform nature, 'having been originally breathed by the Creator.'[10]

The book's other essays followed a similar logic, though they focused on different aspects of orthodox Christianity. Several of the pieces dealt with biblical criticism – introducing English readers to a practice already well developed in Germany – and pointed out, in particular, the human element in scripture. For instance, the layman of the group, the Egyptologist Charles Goodwin, criticized the attempt by 'theological geologists' to explain the Creation account in Genesis in light of the past thirty years of geological discoveries. Taken as a whole, the essays challenged orthodox Christianity to face up to scientific and historical evidences and to abandon the lies and half-truths that had been perpetuated over the centuries. The historian Josef Altholz points out that while *Essays and Reviews* was largely a work of negative theology, there was a partly developed positive message in it. The essay by the Regius Professor of Greek at Oxford University, Benjamin Jowett, came closest to expressing that message fully. He made a plea for freedom of scholarship, of biblical criticism, and of science, arguing that only when that freedom had been granted would we be able to 'find that truth which unites reason to faith.'[11]

For Wilberforce, this was no less than heresy, literally. He sought to have the clergymen among the authors indicted for heresy at the ecclesiastical courts; he also shared his criticism of the book with a wider audience in the pages of *Quarterly Review*[12] – the same venue where, six months earlier, he had published his denunciation of Darwin's *Origin*. Wilberforce explained that it was his 'painful duty' to review the work because the logic therein led to 'infidelity, if not atheism.' What is more, Wilberforce wanted to expose the work as verging on atheism because of who the authors were. Their identity also helped Wilberforce explain the work's vast circulation, which was 'far greater than it would naturally have obtained by its mere literary merits.' 'It has been read' he argued,

> because to all it is new and startling – to some delightful, and to others shocking – that men holding such posts should advocate such doctrines; that the clerical head of one of our great schools, recently elected by a body of staid Conservative noblemen and country gentlemen, and a Chaplain in Ordinary to Her Majesty; two professors in our famous University of Oxford, one whom is also tutor of one of our most distinguished colleges; the Vice-Principle of the College at Lampeter for training the clergy of the Principality; and a country clergymen, famed in his day for special efforts on behalf of orthodoxy; – that such as these should be the putters forth of doctrines which seem at least to be altogether incompatible with the Bible and the Christian Faith as the Church of England has hitherto received it – this has been a paradox, so rare and so startling as to wake up for the time the English mind to the distasteful subject of a set of sceptical metaphysical speculations regarding many long-received fundamental truths.

From the very beginning of the review, Wilberforce made it very clear who exactly the authors were. He wanted to hold them accountable for opening a debate on 'many long-received fundamental truths,' and he went on to consider the joint responsibility of all the authors involved.[13]

The pieces in *Essays and Reviews* had been written quite independently of one another. 'The authors of the ensuing Essays are responsible for their respective articles only,' the book's introduction claimed. 'They have been written in entire independence of each other, and without concert or comparison.'[14] Wilberforce would have none of that. An author *could*, Wilberforce explained, 'exonerate himself [from] the joint liability' only by pointing out the difference in 'aim and purpose' between his own article and the others. He would also, of course, have to request that the article be removed from subsequent reprints and editions, thereby truly differentiating the individual's work from the collection. However, Wilberforce argued, no such defence of that nature had been attempted, and he could not envision such a defence succeeding from any of the authors. The reason for that was quite simple and obvious to any reader of the book: the 'same general tone of writing pervades the whole book,' argued Wilberforce, 'the free handling of most sacred subjects, the free insinuation of doubts, the freedom of asserting, the free endeavour to defend some shadowy ghost of Christianity by yielding up all that has hitherto been thought its substance, is everywhere present.' The book has a 'common life and purpose' he continued, that undermines any individuality contained therein. He was especially astonished at Frederick Temple's involvement with the collection: 'The English Church needs in her posts of trust such men as his past career has made us believe Dr. Temple to be.' But Wilberforce would not let Temple off the hook, and while he believed Temple's article to be devoid of the type of 'sophistries or scepticisms as abound throughout the rest,' his article 'contains the germ of their developed errors.' Wilberforce called on Temple, 'with the manly openness which we believe marks his character,' to 'disclaim his agreement with the views with which he is here connected.'[15]

This was clearly not your typical intellectual debate, and Wilberforce made it clear that you were either with the Church of England defending Anglicanism or you were with the Broad Church Movement denouncing established truth in favour of a godless rationalism. He used the same tactic in his battle against evolution, a battle he saw as part of his fight against the Broad Church Movement. Several well-known Broad Church clergymen had come out in support of Darwin's conception of evolution. The most notable of these men was the Reverend Charles Kingsley. Perhaps more stunning was Darwin's reliance on Charles Lyell throughout the *Origin*, which suggested that the well-known and deeply respected Unitarian geologist would support Darwin's evolutionary line of reasoning. In his review of the *Origin*, Wilberforce expressed dismay over Darwin's misuse of Lyell: 'We trust that he [Darwin] is mistaken in believing that he may count Sir C. Lyell as one of his converts.' Lyell 'shows the fallacy of Lamarck's reasoning' in his *Principles of Geology*, and therefore 'by anticipation confutes the whole theory of Mr. Darwin.' Just as he had called on Temple to disassociate himself from the views expressed in *Essays and Reviews*, Wilberforce called on Lyell's 'help ... with that of his brethren [to] put down [Darwin's] flimsy speculation' in much the same way as had been done to 'its twin though less instructed-brother, the "Vestiges of Creation." In so doing they will assuredly provide for the strength and continually growing progress of British science.'[16] Darwin was not a little concerned that Wilberforce's review might weaken Lyell's support, writing to Hooker that the 'concluding pages will make Lyell shake in his shoes.'[17]

It is important to note that science, for Wilberforce, was not an evil tool that could only be used to undermine religious truths. Actually, Wilberforce enjoyed thinking about scientific questions and debates of the day. The editors of the *Quarterly Review* believed him to have enough scientific knowledge to review the *Origin*, though it should be noted that he relied on other scientific authorities such as Richard Owen in denouncing natural selection. He was also a firm supporter of the BAAS, and his fateful encounter with Huxley on 30 June 1860 was only one of the many talks he attended that week. Indeed, contemporary newspaper reports noted his presence at the other panels as well. It was only after the debate of 30 June gained legs – largely through Darwin's network of supporters – that Huxley's encounter with Wilberforce was singled out and that the importance of the 'Oxford Debate' was born.

Wilberforce was separated from the so-called scientific naturalists not just by his refusal to accept evolution. They also differed over what the precise function of science was. For Wilberforce, science was not doing its job if it was overturning religious truth. He followed William Paley and other natural theologians in believing that the function of science was to uncover and illustrate the beautiful works of God's hands. It was the task of the naturalist to discover the world God created, not to undermine the evidences of His handiwork, which for Wilberforce was precisely what the theory of evolution by natural selection sought to do.

While some of Wilberforce's criticisms of the *Origin* were obviously religiously motivated, for the most part he showed a deft hand in critiquing the book on scientific grounds. As John Hedley Brooke points out, Wilberforce skilfully presented two central and related problems with Darwin's theory.[18] The first concerned the analogy that Darwin was making between the selective breeding of domesticated species and his primary mechanism in evolution, natural selection. If man is capable of selectively breeding certain characteristics in domestic animals, Darwin suggested in the *Origin*, imagine the capabilities of 'natural' selection's 'daily and hourly scrutinizing' over the space of entire geological periods.[19] The problem with this logic, for Wilberforce, was that domestic breeders cannot produce new species; and what is more, wild descendents of selectively bred domestic species 'return from the abnormal to the original type.'[20] If anything, the evidence derived from domestic breeding disproved, rather than proved, evolution.

Wilberforce also criticized Darwin's attempt to 'to break down the unanswerable refutation which is given to his theory by the testimony of the rocks.'[21] Here Wilberforce was referring to the lack of transitional forms one would expect to find in the geological record. Darwin's logic in this instance was also unsatisfactory. He had appropriated Lyell's argument that the fossil record is necessarily incomplete; even so, for Wilberforce this weakness in Darwinian theory was another example of an attempt to transform evidence that clearly disproves evolution into evidence that proves it. Darwin himself had to admit that the review was 'uncommonly clever' and that Wilberforce 'picks out with skill all the most conjectural parts [of the book], and brings forwards well all difficulties.'[22]

But Wilberforce was incapable of limiting himself to a discussion of the evidence and logic for Darwin's theory of evolution. After claiming that he had 'no sympathy with those who object to any facts or alleged facts in nature … because they believe them to contradict what … is taught by Revelation' and that 'we cannot … consent to test the truth of natural science by the Word of Revelation,' he went on to say that 'this does not make it less important to point out on scientific grounds scientific errors, when those errors tend to limit God's glory in creation, or to gainsay the revealed relations of that creation to Himself.'[23] In other words, it is important to point out when scientific errors limit God's divine power, as Darwin's surely did. Evolution, for Wilberforce, and in particular the evolution of man,

> is absolutely incompatible not only with single expressions in the word of God on that subject of natural science with which it is not immediately concerned, but … with the whole representation of that moral and spiritual condition of man which is its proper subject matter. Man's derived supremacy over the earth; man's power of articulate speech; man's gift of reason; man's free-will and responsibility; man's fall and man's redemption; the incarnation of the Eternal Son; the indwelling of the Eternal Spirit, – all are equally and utterly irreconcilable with the degrading notion of the brute origin of him who was created in the image of God, and redeemed by the Eternal Son assuming to himself his nature.

Darwin's theory was more than scientifically wrong; his mistakes formed a theory that contradicted 'the revealed relation of creation to its Creator [and] is equally inconsistent with the fullness of His glory.'[24] There were, then, both scientific and religious reasons for rejecting Darwin's evolutionary hypothesis.

Darwin was well aware of the powerful criticisms that would be directed at his conception of evolution by firm believers in natural theology such as Wilberforce. As stated earlier, Darwin was an avid reader of Paley as well as the Bridgewater Treatises. He was also tormented by the implications of his evolutionary theory, and not just because transmutation necessarily engendered godless thoughts. Darwin's conception of transmutation had replaced the natural theologians' God with Nature, thereby undermining the purpose of the gentlemanly naturalist whose primary task was to reveal God's existence through His works, the natural world. Darwin's twenty-year hesitation in publishing his theory had much to do with the fact that his theory of evolution seemed to undermine what was generally believed to be the naturalist's principal function.

In other words, Darwin anticipated Wilberforce's criticisms, and not simply because they were clearly those of an orthodox Anglican Churchman. Darwin had spent several years as Robert FitzRoy's companion on the HMS *Beagle,* and their friendship turned sour once the two of them sat down to publish individual volumes of a joint history of the *Beagle*'s many travels. While Darwin's own account failed to discuss transmutation – and Darwin's views on that subject would not be made public for twenty years, on the publication of the *Origin* – his discussion of finches as unique species to the Galapagos could confuse a reader who was trying to mesh such a view with a literal interpretation of Genesis. FitzRoy's rediscovery of an ardent faith in God after sailing around the world with Darwin caused him to interpret Darwin's book as if the evolutionary theory of the *Origin* had been apparent twenty years earlier. As FitzRoy's and Darwin's accounts on the voyage were being published together, FitzRoy had a chance to read Darwin's volume before submitting his own to the publisher. FitzRoy revised what he had already written to counter Darwin on points FitzRoy could see being taken in an evolutionary direction. He also added two completely unnecessary chapters at the end of his narrative that made explicit his literal reading of the Bible in relation to all that he had observed during his travels, going so far as to repudiate the largely accepted facts concerning the slow changes occurring in the earth's history.[25]

While Darwin's view of FitzRoy changed from the beginning of the voyage to the end, and while he could rightly have dismissed FitzRoy's views as coming from the clouded mind of a religious zealot, he did not fail to discern similar views espoused by his wife, Emma. Much like Samuel Wilberforce, Emma's religious beliefs were a product of the evangelical revival of the late eighteenth century. Emma remained devoted to the evangelical faith despite her husband's growing unbelief. As Emma began to perceive in Charles's studies an overturning of certain religious truths, she worried for his mental health. Her early concerns for his scientific inquiries were expressed in a letter to him, and they tell us a great deal about Emma's own religious beliefs as well as her great concern for the health of her husband. She understood that Charles was acting 'conscientiously' when 'trying to learn the truth,' but this fact in itself was sometimes not enough comfort, given the results of his studies. 'Your mind and time are full of the most interesting subjects and thoughts of the most absorbing kind, viz. following up your own discoveries, but which make it very difficult for your to avoid casting out as interpretations other sorts of thoughts which have no relation to what you are pursuing, or to be able to give your whole attention to both sides of the question.' Emma proceeded to discuss the work of Darwin's grandfather, Erasmus, 'whom you have so much affection for, having gone before you.' Did this perhaps make it easier for Charles to deal with 'some of the dread and fear which the feeling of doubt first gives, and which I do not think as unreasonable or superstitious feeling?' Emma was concerned, however, that Charles had been spending too much time pursuing the difficulties of one side of the species origin question and that he had not the 'time to consider and study the chain of difficulties on the other ... May not the habit in scientific pursuits of believing nothing till it is proved, influence your mind too much in other things which cannot be moved in the same way, and which, if true, are likely to be above our comprehension?'[26]

Emma felt it important to put her thoughts down in writing because she wanted Charles to truly understand how she felt – that he was casting his scientific inquiry in directions that 'are likely to be above our comprehension. There is danger in giving up revelation which does not exist on the other side, that is the fear of ingratitude in casting off what has been done for your benefit, as well as for that of the world, and which ought to make you still more careful, perhaps even fearful, lest you should not have taken all the pains you could to judge truly.' Charles's studies were also her affair, she argued: 'Everything that concerns you concerns me.'[27]

Charles often referred to Emma's 'beautiful letter' throughout his correspondences with her. He was only too aware of the metaphysical world that he was overturning by exploring the transmutation of species. Indeed, so was Emma. She made it clear to her husband that in following the logic of his own discoveries it was becoming difficult to 'avoid casting out as interruptions other sorts of thoughts which have no relation to what you are pursuing.' For Emma, transmutation posed an extremely important and practical problem to their life together: it suggested that it would end in meaningless death. On this point, Emma was clear: 'I should be most unhappy if I thought we did not belong to each other for ever.'[28] Emma, in other words, had already raised many of Wilberforce's more evangelical-inspired objections.

While Charles clearly empathized with both Emma's and Wilberforce's concerns, he could not fail to see in his family life the same struggle for existence that he saw in nature, a struggle that was marked by his own invalidism and by the spectre of death that constantly haunted the life of the Darwins. Even for Victorians, for whom dealing with death and sickness was a life's constant, the Darwins experienced far more than their fair share. Charles believed that this was nature's way of selecting out his horrid genes so that they would not be passed on. This was all the solace Charles could muster in the face of losing several children to the incessant war of nature. Emma believed that Charles, having turned away from the comforting hand of God, was needlessly tortured and tormented by thoughts of death. In 1861, Darwin's long-time mentor and friend Henslow died. Despite having lost several of his children by this time, dealing with death for Charles was no easier. 'My heart has often been too full to speak or to take any notice,' Emma wrote to Charles in another 'beautiful letter.' 'I am sure you know I love you well enough to believe I mind your sufferings nearly as much as I should my own and I find the only relief to my mind is to take it as from God's hand, and to try to believe that suffering an illness is meant to help us exalt our minds and to look forward with hope to a future state.'[29] Emma felt that Charles was needlessly tormenting himself. His growing agnosticism left him ill prepared to deal with death, which in his mind could only be meaningless. In the *Origin*, Darwin tried to make death sound routine, a mere function of natural selection. 'When we reflect on this struggle,' argued Darwin, at the end of his chapter on the struggle for existence, where he described the incessant war within and between species, 'we may console ourselves with the full belief, that death is generally prompt, and that the vigorous, the healthy, and the happy survive and multiply.'[30] Obviously he was unable to console himself with such a belief when it came to the death of his family members and friends.

It was, in particular, the death of Darwin's oldest daughter Annie in 1851 that sent Charles down the road to reject Christianity outright.[31] In June of 1850, nine-year-old Annie began to complain of stomach pains similar to those her father suffered. She suffered from severe pain for weeks at a time before it would subside and her recovery appeared immanent. She was not getting better. After one of her relapses, Darwin took her out to Malvern to be treated by his 'water cure.' Dr Gully did what he could with his hydrotherapy but Annie did not improve. After several days of struggling to stay alive, she took her last breath on 23 April 1851. 'Annie's cruel death destroyed Charles's tatters of belief in a moral, just universe,' argue Adrian Desmond and James Moore. Darwin himself would later say 'that this period chimed the final death-knell for his Christianity.'[32]

A confrontation with death marked a turning point in the life of Samuel Wilberforce as well. His wife's family was especially troubled by disease and death. Throughout the 1830s, Emily's close relatives had been struck down by heart disease. Between 1829 and 1836, her mother 'had buried her husband, her two sons, a daughter, and her father-in-law.' Meanwhile, Wilberforce had lost his father, his sister, and a sister-in-law. He and Emily lost two of their children, one born prematurely in 1834 and the other a day old in 1837.[33] To say that Emily and Samuel were preoccupied with death in the 1830s would be an understatement of the highest order.

Emily had been married to Samuel for only thirteen years when her family's history of heart problems caught up with her. She caught a fever in October of 1840, while pregnant, and never recovered, passing away on 10 March 1841. Samuel was crushed. On the day of her death, he wrote in his diary: 'A day of unknown agony to me. Every feeling stunned. Paroxysms of convulsive anguish and no power of looking up through the darkness which had settled on my soul.' Emily's death, he believed, was God's punishment for his sins. This was a difficult lesson for him to understand. That same day, he wrote to his friend Louisa Noel, expressing an earnest 'desire to bow to the will of God; to receive His loving correction meekly ... but I am so utterly crushed that I cannot tell what else I feel or what is real around me. I can hardly pray; I seem only in a muffling, horrible dream.'[34]

By 11 March, a day after Emily's passing, Wilberforce had begun to accept her death and was trying to comprehend the message that God had conveyed by taking her from this world. 'In some degree, yet but little, able to look to God, as the smiter of my soul, for healing,' Wilberforce wrote in his diary. 'Oh, may HE enable me to lead a life more devoted to His glory and my master's work.' Emily's death, he wrote, was meant to 'kill in me all my ambitious desires and earthly purposes, my love of money and power and place, and make me bow meekly to Christ's yoke.' This was a turning point in the life of Samuel Wilberforce.[35]

A week after Emily's death, Wilberforce's diary entry shows that he had finally come to terms with God's message. 'It is a call to a different mode of life' he wrote. What else '*can* the sudden removal [of my happiness] mean than that I am to serve Him in a different way: in a more severe, separate, self-mortifying course?' He came to accept Emily's death as a sign that he must devote himself entirely to the Church: that this was God's will. Defending Christian truth would become Samuel's purpose in life, 'his burden of *desolate* service.' '*Above all*,' he wrote, 'I am called ... to a life of much prayer and communing with God. This stills, settles, strengthens the heart. Give up to God the time before given to her.'[36] There is no doubt that Wilberforce saw his battle against the Broad Church Movement, against the rationalization of faith, and against the evolutionists, as his service to God.

It should not be surprising that both Wilberforce and Darwin went through similar periods of moral anguish following the deaths of loved ones, during which their ingrained religious beliefs were questioned. This was the narrative arc described in so many autobiographies of the Victorian era. Wilberforce and Darwin represent the extreme poles of the many possible responses to the era's crisis of religious doubt. Wilberforce found solace in his religion, much like Darwin's wife; Charles could not do the same. Wilberforce became an outspoken defender of Anglican orthodoxy; Darwin became the proponent of evolution and a symbol of the rise of Victorian unbelief.

Darwin knew that his theory would generate resistance and raise difficult questions. Emma herself had raised the question of the afterlife after Darwin gave her his draft of 1844, to be published in the event that he succumbed to his sickness. This was the type of question that made the concept of evolution deeply personal, one that would necessarily cause great anguish among those who tried to take the theory seriously while attempting to hold on to bedrock Christian truths.

Darwin was no revolutionary, as his own anguish about publishing such a work will attest. When he was forced to publish his mere abstract by the appearance of Wallace's letter describing evolution in almost identical terms, he was very careful to make a place for God in his view of life; his hope was to pre-empt religious detractors such as Wilberforce. He knew he would not be able to convince everyone; he would be content if his book met the reading public with a whimper and grudging consensus rather than with a bang and violent controversy. The *Origin* abounds with religious overtones. Indeed, at times he employed the language of a natural theologian, claiming to be uncovering the work of God. He expressed wonder when observing the natural world. His view of God, however, was of an absent master, one who established the initial laws of the universe and then let the world unfold according to the mechanisms He had created, such as natural selection.

Wilberforce and other orthodox Anglicans saw through Darwin's religious rhetoric; other Christians, particularly those of the Broad Church Movement, found that Darwin's theory of evolution provided them with the precise view of God and Christianity they needed in order to come to terms not only with evolution but also with the recent findings of biblical criticism and geology. Darwin did little to correct any theological views of his theory. He would likely have disagreed with the view that evolution by natural selection was consistent with natural theology, but he approved of American botanist Asa Gray's (1810–1888) collection of essays on the *Origin*, which said just that.[37] Indeed, Darwin was quite willing to let others imbue his work with meaning, be it religious or secular. Others, however, were more confrontational than the American botanist. Soapy Sam was not willing to give evolution the religious space necessary to unite it with the Christian faith. Neither, come to think of it, was 'Darwin's Bulldog,' Thomas Henry Huxley.

3 Thomas Henry Huxley and Richard Owen; or, Darwin's Bulldog and the Queer Fish

> Then HUXLEY and OWEN
> With rivalry glowing
> 'Tis Brain versus Brain
> 'Til one of them's slain
> By Jove! it will make a good match
>
> 'Monkeyana,' *Punch*, 18 May 1861

It may have been Darwin who provided the scientific manifesto for the evolution of species, but it was Huxley who became the theory's chief ideologue and paladin. Indeed, Huxley was willing to take the fight for evolution to the terrain where it belonged: the relationship between humans and apes. Even before the *Origin* hit the bookshops, Huxley was comparing the anatomies of the baboon, the gorilla, and the human and suggesting that the interval separating the gorilla from the man was less than that which separates the gorilla from the baboon.[1] Where Darwin was only willing to imply – in what surely must have been the 'understatement of the nineteenth century'[2] – that his theory could shed light 'on the origin of man and his history,' Huxley made an explicit link between the evolution of species and the evolution of man.[3] Furthermore, he was willing to do so in the most public and hostile of venues.

Yet only a few years before the *Origin* appeared, Huxley was adamantly opposed to any theory that posited the transmutation of *any* species. While he chafed at the religious underpinnings of much of Britain's scientific scholarship, he held firm to a fairly orthodox view of species as individual creations. He believed that no theory had been put forward that was both naturalistic and supported by scientific evidence. However, once he grasped the political and social relevance of Darwinian evolution, as well as its scientific merits, thanks to a few gatherings with Darwin at Down House, he became Darwinism's most outspoken advocate. Indeed, Darwinian evolution became for Huxley just the scientific theory to rally behind in order to break the entire scientific establishment in Britain, which he saw as dominated by an unscientific natural theology buttressed by a patronage system tied strongly to the Anglican Church. In many ways the Oxford debate was Huxley's coming-out party, for there he let it be known that religiously motivated criticisms would no longer be tolerated, at least in his presence.

Huxley was born into a lower-middle-class family in 1825 above a butcher's shop in the small village of Ealing, twelve miles west of London (now a suburb of the city). When he was only ten, his family fell on hard times. His father, George, who taught mathematics at a public school, lost his job when the school went under, thus ending the younger Huxley's two years of formal education. The family was forced to move to Coventry, where George took a job at a bank; but he did not make enough money to send Thomas back to school, given that there were seven other children to raise. It is well known that Huxley's scientific education was mostly self taught. He studied invertebrates and vertebrates, and he taught himself drawing as well as the German language, mastering the various subjects into which he delved.[4]

In the 1830s the Industrial Revolution hit Coventry all at once, choking the city with the pollution of several new steam-powered factories, creating a need for many new doctors. Huxley took one of many apprenticeships over the next few years, spending much time in London's notorious East End. The squalor, filth, and disease he found there haunted him for the rest of his life. He was shocked by the middle class's indifference to such misery; and he looked to the Chartists, who sought to expand the franchise to all (male) British citizens, in an attempt to give voice to the forgotten, the weak, and the miserable. Christianity had clearly failed these people, Huxley believed, and something – anything – needed to take its place for the sake of humanity.[5] But an Engels he did not become.

Huxley sought to pull himself up from the drudgery, hoping that he might one day be able to help others. When he was seventeen, in recompense for all his hard work (he had also taught himself Greek and Latin), he was admitted to Charing Cross Hospital in London. His application to the hospital was a lesson in the centrality of the Anglican Church to English affairs that he would not soon forget. He had to have two clergymen vouch for the respectability of his father and family, not a simple task. He went on to study anatomy and physiology at the University of London.

Because of Huxley's lower-middle-class background, nothing came easy for him, except perhaps learning. On his long road to stardom he saw many lesser lights granted the fast track. He saw family members of wealthy Anglicans who had been at Oxford or Cambridge accorded privileges far beyond what they deserved, and for Huxley this was simply wrong. While he excelled at the University of London, winning the gold medal for anatomy and physiology at the age of twenty, he was in far too much debt to continue with his studies, and he was still too young to qualify for the College of Surgeons. Despite his early success at London, he never qualified for a university degree.[6]

Instead, on advice from a fellow student, Huxley wrote a letter to the Director-General for the Medical Service of the Navy, hoping to receive an appointment. After proving that he was not an Irishman (Huxley assumed that his 'air of modesty' must have made him seem Irish), he 'was in Her Majesty's Service, and entered on the books of Nelson's old ship the *Victory*, for duty at Haslar Hospital.' Not long after that, he was recommended to the captain of the HMS *Rattlesnake* 'as an assistant surgeon who [also] knew something about science.'[7] The purpose of the *Rattlesnake*'s voyage, which commenced on 3 December 1846, was to do survey work in Torres Strait off the northern coast of Australia. Once the ship reached the Southern Hemisphere, Huxley spent as much time as possible studying marine invertebrates. A paper of his sent during his voyage was published by the Royal Society in *Philosophical Transactions*.[8]

It is significant that Huxley cut his scientific teeth on a great voyage in much the same way as Darwin and Hooker before him. Their common experiences then sent them in similar social and scientific directions. However, the three of them differed in social standing. Darwin and Hooker came from well-connected and well-off families. Darwin, in particular, could have lived out his days as a country parson. The difference between Huxley and Darwin is perhaps best exemplified by their respective duties during their respective voyages. Darwin's official title during the *Beagle* voyage was 'gentleman companion' to Captain FitzRoy. He had no official duties while on board, and his social standing meant he was entitled to his own personal assistant, which he had. Contrast this with Huxley, who could go on such a voyage only by having an official duty aboard the ship – in this case 'assistant surgeon.' His capacity as someone who 'knew something about science' was simply a bonus, given that the ship already had an official naturalist on board. Hooker's experience was similar to Huxley's – that is, he was forced to accept a position far subordinate to what merit should have accorded him. It should be noted that the chattering classes were taking notice. On 1 December 1846, *The Lancet* wrote that 'some of the best educated young men in the profession are at present acting as naval assistant-surgeons, and "denizens of the midshipmen's berth" ... On looking over the navy list, we find, that such men as the younger HOOKER, the botanist, and HARRY D. GOODSIR, are assistant-surgeons; and also several graduates and undergraduates of the University of London. Among these, we may mention, T. H. HUXLEY, a medalist of 1845.'[9]

Huxley was willing to go along with this for now, however. Indeed, he quickly accepted the necessary duties that came with his position, though he was no less obsessed with making the voyage worth his while. He believed that if he played the game right, and if his work was noticed, he would be offered full surgeon status or, even better, an academic post. With this thought always on his mind, he poured his energy and enthusiasm into his researches into marine biology and then into getting his findings published.

By the end of the four-year voyage, Huxley had accomplished what he intended: he had become a specialist on creatures that no one else in England had examined. He had studied those delicate surface organisms of the sea – hydrozoans, tunicates, and mollusks. What is more, his work had been noticed. On his return he was immediately hobnobbing with the geological elite at Clunn's Hotel. Taken there by the naturalist Edward Forbes, he met geologists Roderick Murchison and Charles Lyell. He was also introduced to the great comparative anatomist, Richard Owen, and quickly began networking with Owen's well-connected Royal Society circle. Despite his hubris, Huxley was actually surprised when he was elected a Fellow of the Royal Society in 1850. The following year he not only received the Royal Medal but also was elected to the society's council. He was only twenty-six.[10]

Things were clearly looking up for him, except for the fact that no promotion was on offer, and neither was an academic post. Huxley was retained by the Admiralty as an assistant surgeon, but that was not exactly the post he had envisioned for himself during his long years at sea. Yet he was not willing to become the voice of rich patrons, which seemed to be the strategy of many other naturalists of humble origins. 'I am under no one's *patronage*, nor do I ever mean to be,' he wrote to his sister. 'I have never asked, and I never will ask, any man for his help for mere motives of friendship.'[11] Huxley's allegiance was to science, to the truth, and he would rather have lived in poverty than betray his faith in a normative meritocracy.

Despite the success of his scientific work, by March of 1851 Huxley had yet to find what he considered reasonable employment for his skills, and he was clearly becoming disillusioned. 'The difficulties in obtaining a decent position in England in anything like a reasonable time seem to me greater than ever they were,' he wrote his sister. 'To attempt to live by any scientific pursuit is a farce. Nothing but what is absolutely practical will go down in England. A man of science may earn great distinction, but not bread. He will get invitations to all sorts of dinners and conversaziones, but not enough income to pay his cab fare. A man of science in these times is like an Esau who sells his birthright for a mess of pottage.'[12]

In May he had grown even more disillusioned, not just because he had yet to receive an academic post, but also because so few posts existed that paid enough to live on. He complained that 'there is no chance of living by science' in England, given the handful of posts available. 'Owen, who has a European reputation, second only to that of Cuvier, gets as Hunterian Professor £300 a year! which is less than the salary of many a bank clerk. A man who chooses a life of science chooses not a life of poverty, but, so far as I can see, a life of *nothing*, and the art of living upon nothing at all has yet to be discovered.'[13]

The creation of a community of scientists, selected according to merit, paid what they deserved, and tasked with finding unfettered truth, seemed much further away to Huxley in 1851 than it did while he was dissecting jellyfish on the *Rattlesnake* a few years earlier. The employment prospects for naturalists who were not also clergymen was next to nil, and Huxley did not have the support, family or otherwise, to eke out a career on accolades alone. Compounding the problem was Huxley's desire to bring to England his future wife, whom he had met while on the *Rattlesnake* voyage – a desire that could only be fulfilled by finding a decent job. With this in mind, he swallowed his English pride and applied for advertised posts in the colonies, one in Australia and another in Canada. He lost out on both positions to individuals of far less merit, betrayed yet again by the patronage system he so loathed. He was beginning to believe that he had 'deceived myself about devotion to Science and the cultivation of the Intellect,' that it was 'all a sham' instead.[14]

Huxley took out his anger on supposed scientific works that he did not think deserved the authority they had been extended. His frustration saturated the review articles he wrote for the periodical press. But at least most journals *paid* for contributions – a small source of income that Huxley was glad to accept. In 1853 he became a writer for the heavyweight liberal quarterly *Westminster Review*, writing a general column on 'Science' that tended to break all of the assumed rules of gentlemanly conduct that governed periodicals. He felt that his role was to act as a 'scientific jackal to the public,' that it was his duty to expose scientific impostures lest the public be fooled into believing poor science. When the tenth edition of *Vestiges* appeared in 1853, Huxley savaged the book in both the *Westminster Review* and the *British and Foreign Medico-Chirurgical Review*. He argued that *Vestiges* was essentially without evidence of knowledge, without evidence of original research, 'nothing but a confusion of ideas, and an ignorance of the first outlines of speculation.' The book had been written for the purpose of convincing unscientific minds, but there was nothing truly scientific about it, he argued. It was, essentially, 'so much waste paper.' According to Huxley, a work like *Vestiges* was the inevitable result of mixing knowledge with commerce, of making the pursuit of knowledge contingent on the power of patronage.[15]

Huxley's review of *Vestiges* may have been a means to promote his own vision of an idealized scientific meritocracy; even so, he did align himself with the old Anglican guard when it came to transmutation. Huxley even defended the geologist Sedgwick, the conservative Cambridge professor, who would later attack the *Origin*, as an honest scientific worker – this, even though Sedgwick's criticisms of *Vestiges* were clearly motivated by an evangelical Christianity. Little did Huxley know that in only a few years he would be aligning himself with the anonymous author of *Vestiges* against the likes of Sedgwick and the rest of the Anglican elite. Only a few months after Huxley's review of *Vestiges* appeared, he finally received that coveted academic post he had longed for, becoming a professor at the Royal School of Mines in London. The following year he took up another position as naturalist to the Geological Survey. It may have taken longer than it should have, but by 1855 he had finally achieved a coveted academic post and, with it, the freedom he believed he deserved. However, his struggles over the years had made him realize that the British scientific system was broken. He would use his new position and status to help fix it.[16]

During the years of struggle between the *Rattlesnake* voyage and his post at the School of Mines, Huxley often invoked the scientific career of the renowned Richard Owen as evidence that good scientific research was poorly funded. Owen, much like Huxley, came from humble origins. He had had to work very hard for everything he achieved, and like Huxley he had been trained as a surgeon through apprenticeship; also, he had briefly attended a university, the University of Edinburgh. In 1832 he had published *Memoirs on the Pearly Nautilus,* which gained a fair amount of notice within Britain's scientific community, largely because of the brilliant anatomical descriptions it contained. He was well on his way to becoming Britain's foremost comparative anatomist.

In 1834 he published a paper on the kangaroo's method of suckling its offspring. Owen claimed that this was clear evidence of design, and his paper was hailed by natural theologians as yet more proof that nature was a product of divine handiwork.[17] About this time, Owen shifted his focus from marsupials to fossil organisms, and his work became, at least in subject matter, more specific to the organic-origins debate. In 1836, Owen was appointed Hunterian Professor at the Royal College of Surgeons, where he remained until 1856, when he became both superintendent of the Natural History Department of the British Museum and Fullerian Physiology Lecturer at the Royal Institution.

Despite his success, Owen could not take for granted his social status within the scientific elite. There was no question that he was a brilliant comparative anatomist, but for him to succeed he could not rely solely on skill. Indeed, Owen depended on the patronage of Prime Minister Sir Robert Peel and a few other Tory supporters. William Buckland even warned Peel that unless Owen was awarded a pension there would always be a real possibility that Britain's foremost anatomist would be 'obliged to descend to the condition of a Bookseller's Hack.' Owen's scientific career, in other words, was always quite contingent on his maintaining a balance between the quality of his work and the expectations of Tory patrons and an Anglican scientific establishment. Owen was stuck in the situation that Huxley desperately wanted to avoid.[18]

Perhaps Owen saw a bit of himself in the young Huxley, and he took him out to dinners to meet his Royal Society circle when the latter returned from his *Rattlesnake* voyage. Huxley was somewhat surprised at Owen's generosity, given that they had only met once, and briefly, before Huxley went on the *Rattlesnake* voyage. Owen superintended the publication of a paper of Huxley's while he was away, and while he was struggling to find work on his return, it was Owen who wrote a letter to the First Sea Lord that secured an appointment to the commodore's ship *Fisguard,* which then proceeded to grant Huxley six months' leave at half pay, renewable.[19] Huxley was honoured by Owen's help, and respected him in return, believing him to be the best naturalist in Britain in 1851 (Darwin was a close third on his list),[20] though he never quite felt comfortable in the man's presence. 'Owen has been amazingly civil to me,' Huxley wrote to his sister, 'and it was through his writing to the First Lord that I got my present appointment.' However, Huxley continued, he 'is a queer fish, more odd in appearance than ever … and more bland in manner. He is so frightfully polite that I never feel thoroughly at home with him.'[21] This unease would manifest itself in a full-out war of words later in the decade, with Huxley and Owen clashing over a variety of issues, most notably over the way to properly critique *Vestiges*.

In the 1840s, when the accelerating discovery of fossils was putting the question of organic origins at the forefront of scientific discussion, Owen offered a theory that allowed for the continued balance of science and religion while avoiding outright evolution. This was the theory of archetypes, a far more acceptable theory to the Cambridge Christians than outright transmutation. Owen produced a synthesis of the seemingly diametrically opposed work by Cuvier and Étienne Geoffroy Saint-Hilaire. Saint-Hilaire was a Lamarckian; even so, Owen concurred with his emphasis on the unity of organisms and the links among them. But Owen also concurred with Cuvier's belief in the adaptable nature of organisms, and now he was seeking to synthesize these two opposing theories. With regard to vertebrates, at least, it was clear to Owen that the similarities among members of different species are undeniable; but, he continued, adaptability could not by itself explain the differences. Instead, relying a little more on Saint-Hilaire than on Cuvier, Owen postulated a general 'archetype' for vertebrates, suggesting that all vertebrates are variations on an original archetype that never actually exists in the flesh.

In hindsight, it is easy to see how such a view could lead someone like Darwin to take the next logical step and adhere to full-out evolution. Owen, however, began to speak of the archetype as analogous to a Platonic form, as an idealized blueprint of the individual organism, and this was much more palatable to his Anglican patrons. Archetypes were, in other words, God's original patterns from which the earth's species were formed, albeit in an altered state. By postulating a theory of archetypes, Owen could explain variations in species while maintaining a belief in the theory of special creations – that all species are mere deviations from original archetypes. Owen appropriated Cuvier's adaptive force to explain organisms' divergences from general patterns (i.e., Platonic forms) towards more specialized designs. The Cambridge elite, the Sedgwicks and the Whewells, embraced Owen's theory because it allowed them to dovetail the ever-expanding fossil evidence with a belief in supernatural Creation. Owen, it is clear, would have preferred to leave the Platonic philosophy out of his theory, but it became necessary for him to 'go along with the Platonisation of his archetype,' argues Nicolaas Rupke, 'because it placed him in a religiously safe corner and gave legitimacy to his transcendental morphology.'[22]

At this time Huxley was no evolutionist, but he began to openly dissent from Owen's theories. That Huxley and Owen would later clash over Darwinian evolution is fairly well known, but how their relationship evolved from one of mutual respect to one of outright hatred is an important part of this story. Indeed, while Huxley initially felt somewhat sorry for the man who had to play by the rules of political patronage, and while he indeed benefited from Owen's help through his patronage network, his attitude towards Owen turned quite sour once his own position improved and Owen's began to seem all the more pathetic. They were always sure to come into conflict, given the proximity of their research, however, and Huxley seemed to take joy in calling attention to his minor differences with Owen, even as the older man remained by all accounts respectful of Huxley, writing many letters of recommendation for him throughout the 1850s. For instance, while Huxley seems to have initially adopted Owen's archetype theory, he made it clear that he could not adopt it wholeheartedly. The archetype was a useful abstraction but no Platonic form, no Godly design.[23] It is unclear how this shift in emphasis changed the interpretation of the fossil evidence, but Huxley was clearly irritated that Owen felt the need to invoke the image of the Creator.

The first real salvo Huxley lobbed at Owen came in his review of *Vestiges*. Chambers had relied on much of Owen's fossil work to buttress his Lamarckian-inspired theory of transmutation. Anyone who thought that Owen was a closet evolutionist given his theory of archetypes would have been thoroughly convinced after reading Chambers's interpretation of Owen's work. Indeed, Owen had written to the author of *Vestiges* on receiving his complimentary copy (sent by the author), and by all accounts was extremely grateful for it, showing no signs of disgust at the use of his work:

> I beg to offer you my best thanks for the copy of your work … which I have perused with the pleasure and profit that could not fail to be imparted by a summary of the evidences from all the Natural Sciences bearing upon the origin of all Nature, by one who is evidently familiar with the principles of so extensive a range of human knowledge. It is to be presumed that no true searcher after truth can have a prejudiced dislike to conclusions based upon adequate evidence, and the discovery of the general secondary causes concerned in the production of organized beings upon this planet would not only be received with pleasure, but is probably the chief end which the best anatomists and physiologists have in view.[24]

While this letter can be seen as evidence for Owen's closet evolutionism, it is more likely the product of Owen's own pragmatism, which offers us a glimpse into his world of patronage politics. Owen believed the anonymous author to be Sir Richard Vyvyan, a wealthy patron of the sciences and someone whom Owen could not afford to demean in any way.[25] Owen later savaged *Vestiges* when it became clear that his powerful friends were lining up against it and that the author was not who he initially thought. It was precisely this kind of two-faced fakery for the sake of patronage that was bound to upset Huxley, and he sarcastically stepped in 'to save the learned Professor's reputation.'[26]

In his own trashing of *Vestiges*, Huxley engaged in a discussion of fossil fish in order to illustrate the character of a proper science. In this discussion it became clear that his real target was not the anonymous author of *Vestiges* but the Hunterian Professor of Comparative Anatomy. Huxley used Owen's anonymous review of *Vestiges* to criticize a faulty position that Owen had earlier stated – that early fossil fish were 'embryonic.' Huxley, being too clever by half, was using a later, anonymous position of Owen's to put forward his own pro-Lyellian view of the absolute uniformity of the fossil record, denying Owen's previous position of progression – a position Chambers utilized for his theory of evolution. Owen was publicly against Lyell's uniformitarian view, and Huxley's review had the effect of defending Lyell while slaying Owen with his own sword.[27]

Huxley continued to bait Owen throughout the decade, criticizing his theory of parthenogenesis (i.e., asexual reproduction) to the Royal Institution and the Linnaean Society, taking great pleasure in his triumphs. Meanwhile Owen took his lumps while remaining, at least publicly, on good terms with Huxley. The grounds for their *public* falling out developed during Owen's stint as a visiting lecturer at the School of Mines. For some reason Owen began to call himself 'professor' even though he was, by title, 'lecturer.' Huxley took this as an affront to his own professional status and in consequence no longer veiled his dislike of the man and his science.[28]

Just as Huxley was growing irritated with Owen's creationist paleontology, he met Charles Darwin, whose materialistic and Lyellian view of nature was more to his liking. They met in the spring of 1853 at a meeting of the Geological Society. Their mutual interest in sea creatures provided common ground. Darwin, who had recently finished his impressive study of barnacles, found much to discuss with Huxley, who himself was working on sea squirts (i.e., tunicates).[29] Darwin just so happened to have some rare sea squirts back at his house, and he offered to let Huxley examine them on the condition that Huxley review his book on barnacles. 'It would give me *great* pleasure to see my work reviewed by any one so capable as you of praising anything which might deserve praise, and criticising the errors which no doubt it contains,' Darwin wrote to Huxley.[30] 'Rare sea squirts,' in Rebecca Stott's words, 'came with strings attached.'[31]

Huxley responded in the affirmative to Darwin's request and was kind enough to send a paper on the cephalous Mollusca that he had been working on. On reading that paper, Darwin could not help but notice Huxley's devotion to the fixity of species. While the latter had distanced himself from Owen's idealistic conception of the archetype, he admitted to being an archetype man nonetheless – that is, to supporting the idea that there are fixed and absolute boundaries separating each archetype. In his response, Darwin, instead of challenging Huxley, posed subtle but pointed questions that were meant to chip away at the theory of the fixity of species. He did this in his typical self-effacing style, confessing that 'it would be ridiculous in me to make any remark on a subject on which I am so utterly ignorant,' then posing problematic questions that suggested quite the contrary.[32] This strategy had already worked well on Hooker, as we shall see in the next chapter, but Huxley proved a harder nut to crack. Whatever the case, he certainly did not take offence: his review of Darwin's barnacle book was, to say the least, glowing.[33]

As Darwin drew Huxley into his close network, however, he realized just how different Huxley's view of species was from his own. He hated Owen's idea about the progressive nature of the fossil record, yet he felt it could easily be transformed into his genealogical view. Huxley, however, believed that individual species could only be structured as if on a sphere surrounding an archetype. There was no progress between fossil forms; 'it was all geometry,' argue Adrian Desmond and James Moore, 'not genealogy.'[34] Darwin wanted to keep Owen's progression without the archetype; Huxley wanted the archetype without the progression.

What is significant about Huxley's view of species is that he held on to a fixed view of species – albeit an unorthodox one – with none of the religious baggage that hindered so many of his contemporaries from making the evolutionary leap. The role of God in nature was, quite simply, not an issue for Huxley. As with many of the other personalities discussed in this book, his religious views are best gleaned from his conversations within his marriage. At a party in Sydney in 1847, Huxley met Miss Henrietta Ann Heathorn ('Nettie'), and they began a courtship that would continue by letter when Huxley returned to sea. Paul White has argued persuasively that Huxley's correspondence with Nettie did much to shape his early identity as a 'man of science,' providing him with an important medium for articulating his beliefs with regard to his work.[35] For instance, while the *Rattlesnake* was mapping the passage up to the Barrier Reef, he wrote to her in order to further explain a conversation the two of them had had earlier concerning his religious beliefs – a conversation that seems to have been unsettling to both parties. Huxley was worried that he was unable to put across his feelings in a way that she truly understood. He was adamant that he did not deny the importance of 'religious speculations'; quite the contrary – he suggested that in fact such a thing 'may be taken as a key to the whole [of a] man's character.' A man's opinions, however, are not entirely what is important but rather 'the temper and tone of mind which he brings to the inquiry. Opinion is the result of evidence.' Without ever explaining what he believed, he suggested that his beliefs could not extend beyond the evidence. Here we see an early indication of Huxley's agnosticism – a term he himself invented several decades later, in 1869. 'As for my own opinions themselves,' he explained, 'I can only say in Martin Luther's ever famous words, "Hier Steh Ich – Gott helfe mir – Ich kann nicht anders" ["Here I stand – God help me – I cannot do otherwise"].'[36] The same logic that led Huxley to question God's role in nature – that is, evidence was lacking – underpinned his failure to convert to transmutationism, or so he claimed.

Huxley eventually came over to the 'dark side,' but as Desmond and Moore point out, it had as much to do with the direction of Owen's work in the late 1850s as it did with Darwin's many failed attempts to convert him.[37] Even before the *Origin* appeared, gorillas and apes and their relationship to man had become a hot topic of conversation among Britain's literati. Owen was at the centre of these discussions. He had dissected more apes than anyone, and he was always ready to provide evidence to counter any suggestions that man was originally a monkey. In 1858 Owen was the president-elect of the BAAS, and he used his presidential address as a pulpit to prove man's special status. It was the brain that truly separated the man from the ape, argued Owen: a unique lobe (the hippocampus minor) and uniquely large cerebral hemispheres. It was his contention that humans occupy a subclass in the ordering of species, a special place for humans alone.[38]

If Huxley was going to disagree with Owen on this count, he would have trouble staking out territory outside the evolutionist's terrain. Instead of fighting it, Huxley embraced evolution as if he had invented it, contradicting entirely Owen's special human class in a lecture at the Royal Institution in 1858. There he argued that man was a part of nature in the same way as other organisms, and furthermore, that man's mental and moral faculties were fundamentally the same as those of the animal world. He also suggested that there was very little by way of anatomical structure to separate the man from the baboon and the gorilla.[39] 'Now I am quite sure,' he continued, 'that if we had these three creatures fossilized or preserved in spirits for comparison and were quite unprejudiced judges we should at once admit that there is very little greater interval as animals between the Gorilla and the Man than exists between the Gorilla and the Baboon.'[40] Darwin would spend the entire *Origin* avoiding a discussion of man because of the prejudices it would necessarily invoke; Huxley, who had only just begun considering the possibility of transmutation, was leaping head first into the public debate, making it explicitly about men, apes, and morality. Suggesting that man was related to other species could only undermine man's special place in the hierarchy of God's creations – a significant blow to both man's narcissism and the religious beliefs that depended on it. The notion that man was an ancestor of the ape was a polarizing one, for it highlighted so sharply the metaphysical meaning of evolution. At the Oxford debate Wilberforce would exploit that very notion in denouncing evolution. Clearly, Huxley was not afraid to ruffle some feathers.

In June of 1858, Huxley gave his 'Croonian Lecture' at the Royal Society, with Owen and the old guard all present and accounted for. The lecture amounted to a lengthy attack on Owen, in which Huxley criticized his writings on the skull as confused and obscure. He then criticized Owen's archetype (as if the concept was not a central aspect to his own work), arguing that the term with all its Platonic and mystical connotations was 'fundamentally opposed to the spirit of modern science.'[41] If Owen's work is what pushed Huxley towards an evolutionary world view, it is just as likely that Huxley's attacks forced Owen to take a position more consistent with scientific orthodoxy. That aside, Owen certainly could not have taken Huxley's criticisms as anything but personal and vindictive.

By the time the *Origin* appeared, much of the debate over evolution had been exhausted. The geological controversies of the 1830s and the sensation wrought by the Lamarckian evolutionism of *Vestiges* in the 1840s were decades past. By the time Huxley and Owen were battling it out over the brain structures of man and the gorilla, familiar patterns of argument were emerging, with evolution serving as the wedge separating the younger generation of naturalists from the old guard. Yet Owen hesitated when the *Origin* appeared, not sure what to make of its deistic portrayal of evolution, especially given the fact that Darwin, much like Chambers, relied on many of Owen's anatomical observations and seemed to adhere to Owen's argument about the progressive nature of the fossil record. Darwin had attempted to pre-empt any opposition from Owen, initiating long conversations about his findings before the *Origin* appeared, and he had found in Owen a somewhat sympathetic ear. Owen's comments were guarded during these conversations, but it never appeared to Darwin that Owen would be a fierce critic of his book – that is, until Huxley reviewed the *Origin* in *The Times*.

Immediately after the *Origin* was published, Huxley planted himself as an outspoken defender of the theory. 'I am prepared to go to the stake' in support of the *Origin*, he wrote to Darwin after he finished reading the book for the first time. He told Darwin not to worry too much about the inevitable criticism: 'some of your friends at any rate are endowed with an amount of combativeness which ... may stand you in good stead.' He reassured Darwin that 'I am sharpening up my claws and beak in readiness' for the probable battle.[42]

Huxley's public defence of Darwin began with his review of the *Origin* in *The Times* on 26 December 1859. By chance, Huxley had been asked to review the book by one of the newspaper's staff, who had been assigned the review but was out of his depth when it came to scientific matters. Huxley agreed to write it provided that the journalist prefaced the piece with two or three paragraphs of his own.[43] The paper's wide circulation and the review's length (three-and-a-half columns was unusually long for a science book) meant that Huxley's positive view of Darwin's book reached a surprisingly large audience. Darwin was absolutely tickled 'that a good review would land on the breakfast tables of the ruling classes on the day after Christmas.'[44] The Oxbridge elite took notice.

While the review contained an underhanded swipe at him, Owen might still have ignored what was becoming a far too typical tactic in Huxley's writings. Owen, however, could not ignore the letters of complaint he received from powerful members of his patronage circle. 'Sedgwick in Cambridge, Livingstone in the Sudan, the Duke of Argyll in government, and Jeffries Wyman at Harvard, all contacted him' to complain about the book.[45] Owen had no choice but to respond.

First, though, he would have to suffer through yet another of Huxley's reviews of the *Origin*, this one in the *Westminster Review*.[46] The *Times* review had succeeded in marshalling opposition to Darwin's theory; the one in the *Westminster* in effect established the *Origin* as the clarion call for a new secular science now in open battle with the old guard – a battle in which Huxley, Darwin, and the scientific naturalists had history on their side. 'Extinguished theologians lie about the cradle of every science as the strangled snakes beside that of Hercules,' argued Huxley, 'and history records that whenever science and orthodoxy have been fairly opposed, the latter has been forced to retire from the lists, bleeding and crushed if not annihilated; scotched, if not slain.' Not just history, but the 'majesty of Fact is on their side' as well, 'and the elemental forces of Nature are working for them': 'Such men have no fear of traditions however venerable, and no respect for them when they become mischievous and obstructive; but they have better than mere antiquarian business in hand, and if dogmas, which ought to be fossil but are not, are not forced upon their notice, they are too happy to treat them as non-existent.'[47] For Huxley, there were only two choices one could make with regard to the question of species origin: one could be on the side of the facts and adhere to transmutationism, or one could be on the side of ignorance, choosing to believe in the doctrine of special creations and with it the 'verbal hocus-pocus' of Owen's divine archetype.[48]

Perhaps Huxley's most prescient comment came at the very end of his *Westminster* review of the *Origin*, where he suggested that Darwin had done more than anyone in the past thirty years to extend 'the dominion of Science over regions of thought into which she has, as yet, hardly penetrated.'[49] Here, of course, he was suggesting that Darwin had extended science's terrain to territory that had previously been occupied by religious authority. This sort of threat to religious authority was exactly what Darwin wanted to avoid. He certainly enjoyed Huxley's review, but he also felt uneasy about Huxley's tone, which had little scientific value. 'There is a *brilliant* review by Huxley, with capital hits,' Darwin wrote to Lyell, 'but I do not know that he much advances the subject.'[50]

Darwin had little time to revel in Huxley's *Westminster* review, for Owen's anonymous review of the *Origin* hit the streets in April in the *Edinburgh Review*. Huxley and Hooker happened to be spending the weekend with Darwin at Down House when the copy arrived, and the three of them read it together in a combined state of shock, amusement, and anger. Owen did more than rip apart the *Origin* piece by piece; he also challenged the author's credentials, in this way staking out grounds for criticism that all earlier negative reviews had avoided because Darwin by then was well-known and well-liked within the scientific community. Nor was the *Origin* the only victim of that review: Owen used the opportunity to denouce Huxley's recent Royal Institution lecture as well as Hooker's essay on Tasmanian plants, which had been published the previous year. As far as Owen was concerned, these works were all exemplars of the same wrong-headed and unscientific theory of evolution.

Darwin had no doubt about the identity of the review's author. 'I have just read the Edinburgh,' he wrote to Lyell, 'which without a doubt is by Owen. It is extremely malignant, clever and I fear will be very damaging. He is atrociously severe on Huxley's lecture, and very bitter against Hooker. So we three *enjoyed* it together: not that I really enjoyed it, for it made me uncomfortable for one night; but I have got quite over it today.' He could not get over the review's tone, however.

Nor could he understand how Owen could be so duplicitous, pretending in person to appreciate the scientific merits of the work, only to trash it under the veil of anonymity. 'It is painful to be hated in the intense degree with which Owen hates me,' he told Lyell.[51] The battle lines had been sharply drawn.

Darwin, Huxley, and Hooker vowed revenge as they paced the grounds at Down House, perhaps consoled by the fact that the BAAS meeting of that year was only a month away. While Darwin backed out of the meeting, both Huxley and Hooker kept their Oxford date. Huxley planned on making the most of his trip by examining Oxford's crocodile collection in order to further understand the Elgin fossils he was working on. He reserved a room in Christ Church and even had the crocodile specimens delivered there to greet him upon his arrival. He would do his best to get some work done during a week of meetings that he was sure would be even 'duller' than the previous year's.

Owen, too, was preparing to make the most of the week. He was hoping to wrangle further support for a national museum, the idea for which he had introduced during his BAAS presidential address of 1858. In February of 1859 he produced a formal and detailed plan for a national collection of natural history – a vision that would not manifest itself until 1881, when the Natural History Museum finally opened in South Kensington. Little did Owen know that the delay in founding the museum would be caused by a partisan rivalry headed by Huxley, who had favoured such an initiative until Owen began promoting the idea. At the BAAS meeting of 1860, Huxley and Owen would continue their debate about apes and ancestry, but this time their jibes would not come in the veiled form of scientific discourse. This time, their battle would be fully exposed for the public to see.

4 Joseph Dalton Hooker and the Early History of a Great Friendship

> But the making of the 'Origin' is not only the history of science – it is the history of a great friendship. In its fabric the two strands are indissolubly interwoven.
>
> Leonard Huxley,
> *Life and Letters of Sir Joseph Dalton Hooker*

It would be difficult to overemphasize the importance of Darwin's friends – his close circle of supporters – in the development, dissemination, and defence of his evolutionary hypothesis. The figure of Huxley is often invoked in this regard, and for good reason, but Darwin had another equally important friend. Joseph Dalton Hooker was more than an important scientific ally of Darwin's – he was also his closest friend and confidant. Defending Darwinian evolution came quite naturally to Hooker, and not just because he helped Darwin construct the theory. In defending evolution, Hooker was also defending a shared history of conversations, of debate, of mutual respect – he was defending a friendship.

When Darwin first learned of the Oxford debate – the debate where he and his 'book forthwith became the topics of the day' – it was, appropriately enough, in a letter from Hooker. Wilberforce, Hooker explained, 'got up and spouted for half an hour with inimitable spirit ugliness and emptiness and unfairness … ridicul[ing] you badly and Huxley savagely.' Hooker went on to describe Huxley's famous response but did not quite grant it the heroic and mythologized aura it would later receive: 'Huxley answered admirably and turned the tables, but he could not throw his voice over so large an assembly, nor command the audience; and he did not allude to *Sam's* weak points nor put the matter in a form or way that carried the audience.' From Hooker's perspective, Huxley may have scored points against Wilberforce when it came to the latter's quip about ancestry, but Huxley did not actually defend evolution by examining '*Sam's* weak points.' Indeed, where legend would have us believe that Huxley not only rebutted Wilberforce's diatribe but also managed to cause indistinguishable laughter at the bishop's expense, Hooker found Huxley's voice far too quiet for such a large assembly and that Huxley simply did not respond to Wilberforce in 'a form or way that carried the audience.'[1] As far as Hooker was concerned, it was his own response that carried the day.

Hooker described his response to Wilberforce as if reporting on an intellectual boxing match: 'I was cocked up with Sam at my right elbow, and there and then I smashed him amid rounds of applause – I hit him in the wind at the first shot in 10 words taken from his own ugly mouth – and then proceeded to demonstrate in as few more 1 that he could never have read your book and 2 that he was absolutely ignorant of the rudiments of Botanical Science.'

It was Hooker, rather than Huxley, who defended evolution in the face of Wilberforce's religious backlash. The meeting was dissolved immediately after Hooker's reply, as Wilberforce 'was shut up,' leaving Darwin 'master of the field 4 hours after battle.' Even Huxley, 'who had borne all the previous brunt of the battle and who never before (thank God) praised me to my face,' Hooker claimed, 'told me it was splendid, and that he did not know before what stuff I was made of.'[2] Hooker was clearly gloating, but he was also letting Darwin know that his ideas were in good hands despite his absence, that his close friends would heroically defend his theory even against the power of the Church. Paleontologist and friend Hugh Falconer would later write to Darwin about the debate: 'Your interests I assure you were most tenderly watched over by your devoted Elèves.'[3] Darwin must have been pleased that his friendship with Huxley and Hooker was so clearly paying dividends.

Darwin would not have been surprised that it was Hooker who meticulously unravelled Wilberforce's criticism of Darwinian evolution. Hooker, unlike Huxley, was no late convert to transmutation. Hooker and Darwin had a friendship dating back to 1843 based on mutual interests in the botanical sciences. Their correspondence is massive, and it is clear from their early letters, in which they shared observable phenomena and other data, that they deeply respected each other's scientific abilities. Darwin shared with Hooker his early considerations of evolutionary theory; in other words, Hooker was the first scientific figure whom Darwin trusted enough to confide in. In a letter to Hooker dated 11 January 1844, a full fifteen years before the eventual publication of the *Origin*, Darwin hesitantly described the tenuous conclusion he had reached regarding the possibility of the evolution of species. He trusted Hooker with his heretical theory even though he felt as if he were 'confessing a murder.'[4]

Little did Darwin know at the time that Hooker would become more than a casual friend. Hooker was eight years younger than Darwin and was extremely excited to have made his acquaintance and become a correspondent. In 1839, at the age of twenty-one, Hooker was voyaging to the Antarctic on the HMS *Erebus* as the ship's salaried surgeon. Much as Darwin had done on his voyage to the Galapagos, Hooker had used his journey for research purposes. Unlike Darwin, however, it was not Lyell's work that he read during the long days and nights at sea. Hooker had been given a copy of Darwin's *Journal of Researches* as a parting gift from Charles Lyell, and he had devoured the work on his long trip. Hooker had already been reading the proofs to the work, which he kept under his pillow so that he could read about Darwin's voyage in the mornings. He wrote to his mother during his trip that the book was 'not only indispensable but a delightful companion and guide.'[5] He later remembered that 'no more instructive and inspiring work occupied the bookshelf of my narrow quarters throughout the voyage.'[6]

Youthful admiration meant that Hooker would have been very sympathetic to any new idea Darwin presented him. Knowing none of this, Darwin tenuously presented his theory to Hooker in his typical self-effacing style. Darwin admitted to being 'engaged in a very presumptuous work,' one that must seem 'very foolish' to just about everyone. As if practising for what would become the first few famous lines of the *Origin,* Darwin suggested that he 'was so struck with the distribution of the Galapagos organisms' and 'with the character of the American fossil mammifers' that he was 'determined to collect blindly every sort of fact, which could bear any way on what are species.' The evidence, Darwin claimed, led him to be 'almost convinced (quite contrary to [the] opinion I started with) that species are not (it is like confessing a murder) immutable.' Darwin admitted that he was worried Hooker would 'groan' in response and 'think to yourself "on what a man have I been wasting my time in writing to."'[7] Darwin would have to wait several weeks for Hooker's response, but when it came he was more than delighted.

'You take so much notice of me,' Hooker began his letter, dropping his previously guarded admiration, 'that I am almost afraid of saying too much, and of destroying the illusory character you give of my little notes.' Hooker spent the larger part of his long letter discussing other questions raised by his and Darwin's correspondence, but finally responded to Darwin's confession regarding transmutation. While Hooker by no means embraced Darwin's hypothesis about the mutability of species, he also let it be known that he was willing to be persuaded. 'I shall be delighted to hear how you think that this [gradual] change [of species] may have taken place,' he wrote, 'as no presently conceived opinions satisfy me on the subject.'[8]

While Darwin was delighted at the prospect of converting a fellow naturalist to his evolutionary world view as well as having scientific discussions pertaining to his now not so hidden secret, he must have also been aware of Hooker's religious background, a background not unlike his own wife's. Hooker had been born into a devoutly evangelical family, and there is little evidence that he dissented from this religiosity. In his letters to and from his parents while he was away on his various trips of discovery, he referred to his weekly attendance at church. He knew that his parents worried about his religious observance in their absence, and he was always quick to reassure them. While the *Erebus* was preparing to set sail from Chatham, Joseph was visited by his father William, who wanted to ensure that his young Joseph was being properly looked after in the rough-and-tumble world of sea exploration. After returning home, William wrote to Joseph on 27 August 1839 about his concerns: 'I could have wished you had some zealous Natural History companions to keep up the zest of the thing, and though I think very favourably of most of your companions, I could have wished to have witnessed their conversation taking a most scientific and sober turn. Above all I should have liked to have seen them pay more respect to the Sabbath. Do so, my dear Boy, and carry something of the Sabbath into the week and I am sure you will be a happier man for it.'

This indicates that Joseph's father considered science and religion together to be of utmost importance in Joseph's life. Scientific discussion was not viewed as a hindrance to worship, but rather as complementary to it. William was worried that Joseph's shipmates would not be able to provide the suitable scientific and religious environment for Joseph's voyage. William asked his son to observe the Sabbath whether or not it was observed by his shipmates: 'I am sure you will be a happier man for it.'[9]

If religion was the dominant force in the Hooker household, science was a close second. Darwin's childhood interest in beetles and turtles was actually viewed as a distraction by his father from other, more important educational opportunities such as medicine or law; by contrast, Hooker's childhood fascination with moss and other floral specimen was carefully cultivated by his father, who was Britain's premier botanist in the early nineteenth century. While Joseph was growing up, William was the Regius Professor of Botany at the University of Glasgow; in 1841 he was appointed the director of the Royal Botanical Gardens at Kew. Joseph was keen to follow his father's footsteps as a botanist, and he grew up in an environment that allowed for his wish to flourish with little effort. In the words of Leonard Huxley, Joseph Hooker 'did not so much learn botany as grow up in it.'[10]

William did more than simply provide an intellectual environment conducive to his son's botanical interests; he also knew many influential figures who kept him abreast of any scientific ventures taken up by government. This was how Joseph became involved with James Clark Ross's expedition to the Antarctic. Ross was a celebrated explorer and had discovered the North Magnetic Pole. This new expedition in 1839, the so-called Magnetic Crusade, was to do the same for the South Magnetic Pole (no one knew, at this time, that Antarctica was a continent).[11] Joseph's credentials as a naturalist were yet to be established, however, and the only position he could have taken up on the ship was that of the ship's assistant surgeon. To that end, Joseph temporarily halted his study of natural history in order to prepare for medical exams – a decision that, ironically, was meant to further his ambitions as a naturalist.

Hooker, of course, was keen on becoming the ship's naturalist, the position that he believed Darwin had held while aboard the *Beagle*. (Darwin, of course, had been FitzRoy's 'personal companion.') He let it be known that this was the position he preferred, but Ross believed that such a position could only go to a person 'perfectly well acquainted with every branch of Nat. Hist., and must be well known in the world beforehand, *such a person as Mr. Darwin*.' Hooker rightly responded: 'What was Mr. D. before he went out? He, I daresay, knew his subject better than I now do, but did the world know of him? The voyage with FitzRoy was the making of him (as I had hoped this exped. would me).'[12] This line of reasoning did not persuade Ross that Hooker was fit for such a position. Hooker would have to be happy that he was given time and accommodations to partake in his naturalist ambitions while on the voyage, even if he could only do so when his duties as assistant surgeon were complete.[13] As it turned out, Hooker was lucky that his ship was relatively healthy and required little attention from the assistant surgeon; the *Erebus'* companion ship, the *Terror*, was not so lucky. In July of 1839, Hooker reported that there were five men sick aboard the *Terror* for every one sick aboard the *Erebus*, and this was before the ships had even set out on the voyage.[14]

Ross's Antarctic voyage took Hooker on three expeditions to the South: from Tasmania and back in 1840 and 1841; from New Zealand and back to the Falkland Islands in 1841 and 1842; and from the Falkland Islands back to Cape Horn in 1842 and 1843. Hooker was in England again on 4 September 1843, having spent a full four years at sea. The voyage was a success for both Ross and Hooker. Ross made several discoveries during the voyage, including these: the Ross Sea, Victoria Land, the Great Ice Barrier, and two volcanoes, which he named after the expedition's two ships, Mount Erebus and Mount Terror.

For his part, Hooker had collected an immense botanical collection from the Antarctic, New Zealand, and Tasmania – indeed, it would take him fifteen years to analyse it and publish his results. Darwin had already met Hooker, having encountered him on Trafalgar Square in 1839, accompanied by one of his *Beagle* shipmates; and he had closely followed Hooker's *Erebus* voyage by reading letters the younger man was sending to Lyell. Darwin wrote to William Hooker on 12 March 1843 and let it be known that when Joseph returned, his plant collection from the *Beagle* voyage would 'be joyfully laid at [Joseph's] disposal.'[15] It was during discussions between Darwin and Hooker about the collection that Darwin confessed to his belief in transmutation.

It is clear that Darwin's admission to Hooker that he believed in the transmutation of species was not made on a whim. Hooker had just completed an immense voyage and was a fount of knowledge gained from observing exotic and remote places. Darwin was no botanical specialist; he relied on his various correspondents to provide him with much-needed answers to his growing list of questions about floral species. Because of his interest in speciation, Darwin needed to establish the geographical distribution of plant life, and this was precisely the type of knowledge that Hooker would be able to provide, given his recent voyage.

Darwin was also aware of the prejudices his views might invoke, and he was strategic in disclosing his theory. He had already approached others in a more subtle manner so as to avoid charges of transmutationism. For example, he had approached his former mentor, Henslow, as well as Lyell, expressing an interest in variations and distributions of species. Both were aware that he was working on such a project, but Darwin left them guessing what he actually thought about evolution. Despite Darwin's hints to Lyell about the possible transmutation of species, the latter emphasized emphatically in successive editions of his *Principles of Geology* that there was little evidence to support a theory of evolution. After 1838, Darwin was largely silent to Lyell with regard to his work on species, despite their continued correspondence. Darwin certainly would have liked the support of the eminent Lyell, but he was not about to risk losing Lyell as a friend by engaging in what might have turned into a divisive debate.[16]

Hooker was without the baggage of a Lyell or a Henslow; he did not have a reputation to place at risk by considering transmutation; and he was of a much younger generation and thus less encumbered by the debates and constraints that were central to his father's world. His recent botanical work made him very useful to Darwin; but it is likely that his very youth, along with his lack of baggage associated with the scientific establishment, is what led to Darwin's open confession. And so began one of the most important friendships in the modern history of science. It is important to understand that the history of the *Origin* is not just a history of science – it is also the history of this great friendship, to paraphrase Hooker's biographer, Leonard Huxley.[17] We cannot truly understand one without the other.

Hooker made many visits to Darwin's home in the village of Downe, where the men discussed a number of matters, not all of them scientific. Hooker would recall in later years that these excursions to Down House were primarily a chance for Darwin to 'pump' him for information: 'It was an established rule that he every day pumped me, as he called it, for half an hour or so after breakfast in his study, when he first brought out a heap of slips with questions botanical, geographical, &c., for me to answer. And concluded by telling me of the progress he had made in his own work, asking my opinion on various points.'[18] This early-morning routine was usually followed by afternoon walks, where the conversation was much less formal, covering 'foreign lands and seas, old friends, old books, and things far off to both mind and eye.'[19]

In the fifteen years between Darwin's confession to Hooker and the publication of the *Origin*, Hooker was indispensable to Darwin, who developed evolutionary theory with his help. Darwin was especially interested in Hooker's understanding of the variation and geographical distribution of floral species. This Darwin was able to glean from their conversations and written correspondence, as well as from Hooker's published work, especially his description of New Zealand's flora in his *Flora Novae-Zelandiae* (1853), which Darwin thought 'will be very useful to me, whenever I undertake my volume.'[20] (Here Darwin was referring to his 'big species book,' which he had been working on throughout the late 1850s.)

Darwin also benefited from the fact that Hooker continued exploring the flora of faraway places, refusing to rest his scientific credentials on the basis of one expedition, unlike Darwin, whose *Beagle* voyage was his first and last trip abroad. In November of 1847, Hooker went travelling in India, Nepal, Sikkim, and Tibet, returning to England only in 1851. In 1854 the Royal Society awarded him its Royal Medal for his botanical researches in India and the Antarctic. As stated by the President of the Royal Society, this was in part for Hooker's work on 'the origin and distribution of species.'[21] Indeed, Hooker's work was already contributing to a burgeoning field that would have to wait another five years for Darwin's immense contribution. By 1855 Hooker's scientific credentials were more than established. That year he was appointed assistant to his father, the Director of the Royal Botanical Gardens at Kew, with the understanding that he would rise to his father's post when the time came. And this he did, ten years later, when his father died.

Darwin and Hooker had a wonderful respect for each other's work, and they were always prepared to give each other credit where it was due, which it often was. Hooker dedicated his *Himalayan Journals* (1854) to Darwin. In the *Origin*, meanwhile, Darwin could not 'let [the] opportunity pass without expressing my deep obligations to Dr. Hooker, who for the last fifteen years has aided me in every possible way by his large stores of knowledge and his excellent judgment.'[22] Having read a draft of the *Origin*, Hooker admitted to Darwin that his friend's speculations regarding natural selection had been a 'jam-pot' for his own work. Darwin responded to this wonderful praise by thanking Hooker for the constant support he had provided over the years for such a contemptible theory: 'The truth is I have so accustomed myself, partly from being quizzed by my non-naturalist relations, to expect opposition and even contempt, that I forgot for the moment that you are the one living soul from whom I have constantly received sympathy. Believe that I never forget even for a minute how much assistance I have received from you.'[23]

Despite their friendship and mutual admiration, it took many years of debates and conversations before Darwin arrived at a convincing theory of transmutation, at least as far as Hooker was concerned. From the moment Darwin confessed his 'murder,' Hooker was open to discussions about the reasons for the distribution and variations of species, but he was sceptical about Darwin's specific theory and pushed him over the years to consider a wide range of criticisms as well as other possible hypotheses. Hooker's 'Introductory Essay' to his *Flora Novae-Zelandiae* is an excellent example of his scepticism towards an evolutionary position but also of his willingness to consider the evidence in its favour. He admitted that in terms of methodology he had to remain an anti-evolutionist because he was required to give some permanence to descriptions of his floral species. Hooker went on to discuss the usual critique levelled against evolution, but he also shifted some way towards the evolutionary point of view. He admitted that an isolated population of a given species could change owing to local conditions to the point where it would *appear* to be a new species. It was also clear to him that plants could thrive in different parts of the world even after being transported from different climates, which suggested that organisms had been distributed around the globe by natural means. Hooker undermined much of the evidence for the special creation of individual species, and while he would not admit to a transmutationary origin of species, he was certainly advocating an ongoing natural one.[24]

Darwin and Hooker seemed to disagree most strongly over the very conception of species as well as the implications for the concept once evolution was thrown into the mix. Hooker's primary field, botany, had long been subjected to a constant naming and renaming of the same floral species, to the point where one particular species could have several names linking it to many different and likely unrelated ones. Because of this, Hooker was a 'lumper,' believing it better to assume a wide definition of a particular species and to accept a small percentage of lumping 'close species' together as one, in order to avoid the problem of naming the same species, for example, ten times.[25] Darwin was sympathetic to Hooker's rather loose and pragmatic conception of species, and had he been able to produce a more suitable and communicable concept to take its place, he certainly would have done so. The species concept, however, was here to stay, and as John Beatty has shown, Darwin shaped the definition of the concept to suit his evolutionary theory without changing the use of the term, by using it 'in accordance with *examples* of its referential use by members of his naturalist community.'[26]

Hooker's resistance to transmutation, then, was largely based on methodological rather than religious considerations, even though he was a fairly devout evangelical Anglican. Darwin and Hooker rarely discussed the religious implications of evolution; Hooker, however, did venture along such avenues of discussion with Darwin's other great botanical correspondent, Asa Gray. Darwin had let Gray in on his and Hooker's secret on 7 July 1857, and Gray and Hooker exchanged several letters in which they discussed the possibility of evolution. On 27 August 1857, Hooker made it clear that he did not consider evolution to be inconsistent with 'the most exalted conception of the Deity.' He argued that he could not see why it was 'honorable to the Deity to have made species by separate impulses [and yet] dishonorable to have made them by transmutation – in neither case do we approach the mystery of species.'[27] Here Hooker was presenting a heterodox view of God – that is, as a Deity who acts solely through natural laws instead of constantly intervening in the world, enacting miracles or creating species. This, of course, is the deistic portrayal of evolution found in the *Origin*.

Whether Hooker influenced Darwin's use of religious metaphors throughout the *Origin*, we do not know, but what is clear is that probably more than any other person, Hooker helped shape Darwin's conception of evolution by natural selection, and that by the time the *Origin* was published, intense conversations concerning evolution had been part of Hooker's life for fifteen years. Indeed, Darwin worried that he had forced Hooker down a path of inquiry that had stifled rather than stimulated his friend's scientific growth. By the time the *Origin* appeared in print, however, Hooker had already published a work that made clear his own shift from a static and unchanging view of floral species towards acceptance of complete transmutationism. Darwin feared that he had stolen from Hooker rather than the other way around: 'I keep on feeling (even while differing most from you) just as if I were stealing from you, so much do I owe your writings and conversation: so much more than mere acknowledgements show.' Darwin was convinced that he was the one who had benefited from their close friendship: 'You will never convince me that I do not owe you *ten* times as much as you can owe me.'[28] In many ways, evolution had become just as much Hooker's theory as Darwin's. Richard Owen certainly believed this to be the case when he reviewed the *Origin* and lumped Hooker's recent work together with that of Darwin.

Hooker may have been insulted by Owen's review (see the previous chapter); but he was also proud to have been noticed as a contributor to Darwin's work even as it was being picked apart. Whenever he discussed Owen's review of the *Origin,* he always mentioned Owen's 'snub' of his own work on evolution.[29] He was certainly more than a mere 'disciple' of Darwin's, as Owen put it, and in Darwin's by now typical absence, many began to look to Hooker to explain the theory of evolution by natural selection.

In the wake of Owen's review, Hooker had written to his old friend and fellow botanist William Henry Harvey about the 'middle way … Owen is hedging for … in his review of Darwin and snub of me in [the] *Edinburgh Review.*' This 'middle way' was Owen's muddled theory of evolution via the archetype, for which Hooker 'failed to find any cumulative support in facts.'[30] Harvey, it would seem, was not convinced by natural selection and was suggesting to him Owen's possible but improbable alternative. 'Give time, abate prejudice, and let your ideas clarify, which they will assuredly do in time,' Hooker told Harvey in his attempt to convince him of Darwin's view of species. 'Remember that I was aware of Darwin's views *fourteen years* before I adopted them, and I have done so *solely* and *entirely* from an independent study of plants themselves.'[31] Hooker was willing to consider alternatives to Darwin's theory but only if such theories meshed with his own observed knowledge of plant life, and he had already interpreted such evidence in the framework of Darwinian evolution.

With thoughts of Owen's review on his mind as well as his attempts to convert Harvey, Hooker travelled to Oxford for the annual BAAS meeting. He was more than prepared to defend a theory that had become his own, but like Huxley he doubted that anything useful could be produced during the proceedings. To make matters worse, his wife was unable to accompany him and, of course, his closest friend was too sick to attend, which made the trip much less pleasurable. 'I came here on Thursday afternoon and immediately fell into a lengthened reverie,' Hooker wrote to Darwin from Oxford. 'Without you and my wife I was as dull as ditch water and crept about the once familiar streets feeling like a fish out of water.' '[I] amused myself,' he continued, 'with the College buildings and alternate[d] sleeps in the sleepy gardens and rejoiced in my indolence.' He even 'swore' that he 'would not go near a Section and did not for two days.' But on Saturday he finally stumbled into a session only to find Darwin's book the topic at hand. The president of the 'section,' knowing that Hooker had considered the theory with reference to the botanical sciences, asked for his opinion on the subject.[32] Perhaps something useful would come of the association meeting after all.

5 The Oxford Debate

> The theory of Dr. Darwin … gave rise to the hottest of all debates.
>
> *The Press*, 7 July 1860

> Now that I hear that you and Huxley will fight publicly (which I am sure I never could do) I fully believe that our cause will in the long run prevail. I am glad that I was not in Oxford, for I should have been overwhelmed, with my stomach in its present state.
>
> Charles Darwin to Joseph Hooker, 2 July 1860

By the time of the BAAS meeting of 1860, it was clear that Darwinian evolution had deeply split the British scientific community. The great division in geology was largely reproduced in the debate surrounding Darwinism, given Darwin's clear debt to the uniformitarianism of Lyell and the unresolved implications for Anglican theology that such a view entailed. However, the debate surrounding evolution itself had in some ways been exhausted over the previous few decades in response to the *Vestiges*. This is not to say that the *Origin* did not invoke heated battles and controversy; as we will see, it did. However, the debate was somewhat of a slow burn, though once the fire started, it burned brightly.

Perhaps the most obvious difference between the response to *Vestiges* and the response to the *Origin* is reflected in the BAAS Presidential Addresses of 1845 and 1860. Herschel in his 1845 address had called on the scientific brotherhood to demolish the heresy expressed in *Vestiges;* in the 1860 address there was no mention of the *Origin* at all. But on Thursday, 28 June, and again on Saturday, 30 June, Darwin's book would take centre stage as two separate papers promised to comment on the *Origin of Species*.

On Thursday the Oxford Professor of Botany Charles Daubeny presented a paper 'On the Final Causes of Sexuality of Plants with particular reference to Mr. Darwin's Work "On the Origin of Species by Natural Selection,"' in which he offered lukewarm support for Darwin's theory. Huxley and Owen were both present. After Daubeny's talk, Huxley was called on to voice his opinion. He declined, stating that a public venue such as the British Association was not the place to discuss Darwin's theory, given that 'sentiment would unduly interfere with intellect.' He also stated bluntly that 'Dr. Daubeny had brought forth nothing new to demand or require remark.' At this point, Huxley was still of the opinion that little of scientific value could come out of the association meetings – an opinion that would change over the course of the long weekend. After a few speakers, the floor was granted to Owen, who took the opportunity to describe some facts that would be useful to the public in order to come to terms with the hypothesis of Mr Darwin. He described the vast differences between the brain of a man and the brain of a gorilla, in contradistinction to Huxley's opinion on the matter. Unable to let Owen have the last word on the subject, Huxley jumped to his feet to express his diametrical opposition. He told the audience that he did not have the time or the energy to go into all of the physiological evidence in favour of the similarities between man and ape, but he pointed out that the opposition to such a belief was based on the fear that man might lose his privileged place among the earth's species rather than on an understanding of the facts. Indeed, he continued, men had nothing to fear if it were to be shown that apes were their ancestors.[1]

This minor skirmish was merely the appetizer for Saturday's main course, had the participants but known it. A seemingly harmless paper on the intellectual history of Europe was to be given on Saturday in Section D, and few were excited about it. Indeed, it seems that no one wanted to attend the reading of John William Draper's (1811–1882) paper 'On the Intellectual Development of Europe'; and had the title ended at 'Europe,' the attendance surely would have been low, and the 1860 Oxford debate an obscure footnote in the history of science. But like every good Victorian title, Draper's went on to further describe the paper following a comma – that the intellectual development of Europe would be 'considered with Reference to the Views of Mr. Darwin and others, that the Progression of Organisms is determined by laws.' Making the paper a comment on Darwin's recently published *Origin* surely increased the attendance. But even then, without the foreshadowing provided by whispers that the Bishop of Oxford was planning on using the venue to 'smash Darwin,' Darwin's main defenders would probably have skipped the talk entirely.[2] Huxley was tired and wanted to rejoin his wife at her brother-in-law's house near Reading. He was persuaded to attend Draper's talk by of all people Robert Chambers, who was trying to drum up support to defend Darwin in the face of certain attack.[3] Hooker, as well, had little interest in attending the session, but becoming 'equally bored of doing nothing,' decided at the last minute to go and hear what Wilberforce had to say.[4]

The meeting itself took take place in the newly built Museum of Natural History. The museum was the product of more than a decade of struggle by a group trying to enhance the status of scientific studies at Oxford, which had long ago been overtaken in that regard by its rival Cambridge. Wilberforce had been one of the big names supporting the building of the museum, which was to embody natural theology and to be a place where students could 'contemplate authentic physical wonders of Nature and the greatness of godly design.'[5] More than seven hundred people showed up hoping to hear the Bishop of Oxford speak. The attendees were an odd mixture of dons, undergraduate students, clergy, naturalists, and women. They filled the lecture room until people were spilling out into the hall. The organizers sheepishly moved the talk to the yet unfinished library, a long, narrow room on the building's west side. That move must have given the talk a more informal atmosphere, as the library was not properly equipped for a formal talk of this nature. The president of the session was, appropriately enough, Henslow, who sat next to Draper along the east side of the room between the two main doors. In the audience was Hooker, sitting next to fellow Darwinian John Lubbock. Huxley was seated right next to the elderly Sir Benjamin Brodie, former surgeon to William IV and Victoria. Darwin's former companion on the *Beagle* voyage, Admiral Robert FitzRoy, was also in attendance. Wilberforce, arriving late, pushed to the front and sat on Henslow's other side, as if he was also scheduled to give a talk.[6]

Even before Draper could begin his talk, he was preceded by several announcements, which irritated an already anxious audience. Mr M'Andrew reported on Dr Kinaban's 'Results of Dredging in Dublin Bay,' while Mr Westwood told a fascinating story of an insect that on account of its anomalous character had been classified in three different groups of the order Insectus. Next up was Dr Daubeny, who was kind enough to invite association members to visit his experimental garden in Oxford. Then Dr Lankester brought attention to the first part of Mr Blackwell's work on British spiders. An anxious audience grew even more restless once they realized they would also have to sit through a lecture by Dr Collingwood 'On the Recurrent Animal Form, and its Significance in Systematic Zoology' – a talk noteworthy for the fact that no one in attendance would be able to recall later that it had been given. Then came Draper's talk, which provided no respite from a long week of talks and debates.[7]

The possibility that had brought the audience to the section – that of an anti-evolutionary diatribe from Wilberforce – kept people in their places through Draper's lecture, which was, by all accounts, hopelessly boring. The future historian John Green could not wait for Draper – who at the time was professor of chemistry at New York University – to cease his 'hour and a half of nasal Yankeeism.'[8] Hooker was equally dismissive of Draper, later referring to him as a 'yankee donkey' and to his talk as 'all a pie of Herb[ert] Spencer and [Henry Thomas] Buckle without the seasoning of either.' Hooker would later declare that he would not normally have sat through such 'flatulent stuff' but 'hearing that Soapy Sam was to answer I waited to hear the end.'[9]

Yet Draper's paper was a wonderful example of how quickly Darwinian evolution was being incorporated into already existing areas of research. Hooker's comments about Draper's similarity to Spencer and Buckle were a case in point. Both Spencer and Buckle had written about the progress of European knowledge; and Buckle's *History of Civilization in England,* published in 1857, had already caused a sensation as well as furious debate about the relationship between science and the human past. Draper, it would seem, was merely providing an explicitly Darwinian twist to ideas that were already quite well known and, implicitly at least, evolutionary. Indeed, Michael Ruse has argued persuasively that evolution was clearly derived from the theories of progress of the eighteenth and nineteenth centuries.[10] Draper had already written the book on which his talk was based, though it would remain unpublished until 1863;[11] but it was easy enough to replace 'progress' with 'evolution' in order to make his research relevant to a post-*Origin* world. Draper himself would claim that very few changes had been necessary in order for his historical theory to incorporate

Darwinian evolution. By all accounts, however, his appropriation of Darwin was not a crowd pleaser.

Draper, who had left Britain for the United States in 1832 at the age of twenty-one, was lampooned by the hostile audience for his accent, however slight it may have been. 'I can still hear the American accent of Dr. Draper's opening address,' reminisced Isabel Sidgwick, '"Air we a fortuitous concourse of atoms?"'[12]

As if cribbing from Buckle rather than Darwin, Draper argued that the progress of civilization was determined not by chance but by observable laws. By making an analogy between the history of nations and the history of organisms, Draper argued that the intellect could only progress under favourable circumstances. He then gave a lengthy description of the physiology of man and of how certain physiological characteristics impact the intellect. In all, the talk was about an hour and a half, and none of the key combatants seemed ready to fight, as many were still rubbing their eyes when Henslow opened the floor to 'valid arguments.'

Reverend Cresswell began the discussion by denying that 'any parallel could be drawn between the intellectual progress of man and the physical development of the lower animals.' Sir Benjamin Brodie did not even direct his comments towards Draper's paper, stating bluntly that he simply 'could not subscribe to the thesis of Mr. Darwin.' Darwin, argued Brodie, had not demonstrated man's primordial origin, and, furthermore, the fact of man's unique self-consciousness could not have originated in the lower organisms. 'This power of man was identical with the Divine Intelligence,' Brodie declared, 'and to suppose that this could originate with matter, involved the absurdity of supposing the source of Divine power dependent on the arrangement of matter.'[13] Richard Greswell also spoke against Darwin. As did a Mr Dingle, who actually jumped up to use the blackboard: 'Let this point A be man, and let that point B be the mawnkey.' Cries of 'mawnkey' by the undergraduate contingent shouted Dingle down. Clearly, the audience was getting restless. Henslow continued the proceedings but demanded that the discussion remain on scientific grounds.[14] With several anti-Darwinians having spoken, Henslow turned to Huxley. Huxley declined.[15]

With that, the Bishop of Oxford rose to speak, giving the crowd what they had waited so long for. Heeding the president's warning to keep the discussion on a scientific footing, and having been coached earlier by Owen (who could not make the discussion), Wilberforce likely surprised everyone by scrutinizing Darwin's supposed inductive methodology. By the 'principles of inductive science,' argued Wilberforce, Darwin's theory could not be proven, at least not by the facts Darwin himself had presented.

Where, he asked, are the many intermediary forms that should fill the fossil record? The fact remains that there are none. The fossils of plants and animals and man in the earliest record of the human race are exactly as they are now. Surely we would see changes in species over hundreds of years, despite Darwin's lengthy evolutionary time frame. But we don't. Everywhere is evidence, not of endless changes in species, but of their permanence.

Wilberforce went on to point out that Darwin's great analogy between artificial and natural selection actually worked to *disprove* evolution. All the experiments of breeders had confirmed that one species cannot be mutated into another, no matter how many generations of artificial breeding take place. 'Everywhere sterility attended hybridism,' argued Wilberforce, citing 'the closely-allied forms of the horse and the ass.' In short, argued Wilberforce, after having gone on for more than half an hour, for evolution to be proven, it would have to be observed either in the fossil record or by the action of breeders, and the fact was that both these areas of research had proved the exact opposite. Darwin's book, then, was a work not of science but of philosophy; and as far as philosophy went, Wilberforce believed that Christianity offered a much better philosophy of life than the *Origin*. He shuddered to think of a world where Darwinian evolution would be adopted as a creed, and he worried that Darwin's disciples were doing just that, ignoring both the scientific and philosophical problems with the work.

Before he finished, Wilberforce declared that he was very glad that the 'greatest names in science' had already opposed Darwin's theory, a theory that he 'believed to be opposed to the interests of science and humanity.'[16] With that, Wilberforce turned to Huxley, 'who is about to demolish me,' and asked, 'Was it through his grandfather or his grandmother that he traced his descent from an ape?'[17]

A roar of approval from Wilberforce's sympathizers was heard throughout the museum, as if the crowd had been waiting patiently for this one glorious moment to revel in the bishop's violent poetry. Not many in the crowd realized that Wilberforce was continuing a discussion between Owen and Huxley that had taken place the previous Thursday, that Huxley had already suggested that it did not matter whether or not humans were related to apes, and that this was a humorous follow-up. Such had little immediate relevance.

Despite Wilberforce's apparent victory, Huxley clearly had an ace up his sleeve, for he whispered to Brodie sitting next to him: 'The Lord hath delivered him into mine hands.'[18] Huxley would recount: 'That sagacious old gentleman stared at me as if I had lost my senses. But, in fact, the Bishop had justified the severest retort I could devise … I was careful, however, not to rise to reply, until the meeting called for me – then I let myself go.'[19] Huxley responded that while he had listened very closely to the bishop's speech, he was 'unable to discover either a new fact or a new argument in it – except indeed the question raised as to my personal predilections in the matter of ancestry.' Huxley acted surprised that such a topic would have been brought up during such a serious discussion (as if he had not raised the topic two days earlier) but that he was quite ready to meet the Bishop even on that ground. 'If then,' said Huxley, 'the question is put to me would I rather have a miserable ape for a grandfather or a man highly endowed by nature and possessed of great means and influence and yet who employs those faculties for the mere purpose of introducing ridicule into a grave scientific discussion – I unhesitatingly affirm my preference for the ape.'[20]

Even though most of the attendees were sympathetic to the Bishop, Huxley received as much applause as his enemy if not more. 'The retort was so justly deserved,' the future Cambridge Professor of Economics Henry Fawcett wrote in *Macmillan's Magazine*, 'and so inimitable in its manner, that no one who was present can ever forget the impression it made.'[21] Lady Brewster, wife of the Scottish astronomer David Brewster, fainted and had to be carried out.[22] As far as Huxley was concerned, the day had been won. However, he had hardly said anything in defence of Darwinism. Then FitzRoy rose, Bible in hand, and expressed regret for sailing Darwin around the world and thereby helping him produce a theory in contradiction with the word of God. After Lubbock rose to claim conversion to the Darwinian hypothesis until a better theory came along, Henslow asked Hooker, given his botanical training, to offer his critique of Darwin's theory in light of Wilberforce's comments that all men of science were hostile to it.

Hooker rose and provided a lengthy rebuttal to Wilberforce. In its reportage of the debate, *The Athenaeum* granted the most space to Hooker's response – much more than Huxley received and more than twice the space devoted to Wilberforce's initial critique, which had drawn the ire of Darwin's 'elèves.' *The Athenaeum* reported that

Hooker calmly explained that he had been privy to Darwin's hypothesis for the past fifteen years and had applied Darwin's views to 'botanical investigations of all kinds in the most distant parts of the globe, as well as to the study of some of the largest and most different Floras at home.' His attempts to convert Harvey had clearly paid dividends, as his defence of evolution closely followed the arguments set out in his letters to Harvey. He went on to admit that he had begun the study of natural history under the assumption that 'species were original creations' but that he could no longer support such a view because it could no longer explain the evidence at hand. Hooker disavowed Wilberforce's claim that Darwin's views were being adopted as a creed, arguing instead that evolution by natural selection simply 'offers the most probable explanation of all the phenomena presented by classification, distribution, structure, and development of plants in a state of nature and under civilization.' Hooker was willing to abandon evolution by natural selection, but only if and when a better hypothesis appeared. Until then, Darwinian evolution offered the 'best weapon for future research' at the necessary expense of the doctrine of special creations.

Hooker continued by subtly ridiculing Wilberforce, suggesting that it appeared 'his Lordship' had completely misunderstood Darwin's hypothesis. Wilberforce had intimated that Darwinian evolution had to do with the 'transmutation of existing species into another.' Of course, Darwin had said nothing of the sort, and Hooker 'could not conceive how any one who had read [the *Origin*] could make such a mistake' (which suggested, of course, that Wilberforce had not read the book). This was damaging stuff, and if Huxley got marks for one-upping Wilberforce's quip, it was Hooker who made him look out of his scientific depth.[23]

Once Henslow dissolved the meeting following Hooker's reply, the Darwinians clearly felt that the day was theirs. 'Samuel will think twice before he tries a duel with men of science again,' Huxley wrote to Frederick Dyster, in firm belief that the bishop and his party were 'shut up.' 'I believe I was the most popular man in Oxford for full four and twenty hours afterwards.'[24] Hooker, too, was ecstatic after the 'battle,' claiming that 'I have been congratulated and thanked by the blackest coats and whitest stocks in Oxford ... and plenty of ladies too have flattered'.[25] Darwin was simply glad to have avoided the whole situation: 'I would as soon died as tried to answer the Bishop in such an assembly.'[26]

John Green referred to the whole episode as an 'episcopal defeat.'[27] Another observer believed that Huxley was clearly the 'hero of the day.'[28] The Reverend W.H. Freemantle argued that the end of the meeting left the 'impression that those most capable of estimating the arguments of Darwin in detail saw their way to accept his conclusions.'[29] Indeed, at the very least Huxley and the Darwinians had 'carried an unwilling audience'; in the words of Leonard Huxley, 'instead of being crushed under ridicule,' Darwin's theory surprisingly 'secured a hearing' in the very hostile environment of Oxford.[30] The *Evening Star* reported that the debate itself had made at least one thing clear: that 'the new Darwinian theory, whatever may be its real merits in a scientific point of view, has no small number of supporters amongst the members of the Association.'[31]

Yet Wilberforce and his defenders were no less convinced of their victory. Perhaps this is why Wilberforce showed no ill will towards Huxley in their future encounters.[32] Wilberforce believed that he had 'thoroughly beat' Huxley during their 'long fight.'[33] Balfour Stewart, director of the Kew Observatory and a member of the Royal Society, also thought that 'the Bishop had the best of it.'[34] *John Bull* reported that 'professor Huxley and (to some extent) Hooker were opposed by a powerful phalanx, led by the Bishop of Oxford.' There was no doubt in this reporter's mind that the bishop had carried the day: 'The impression left on the minds of those most competent to judge,' the report continued, 'was that this celebrated theory [Darwinian evolution] had been built on very slight foundations, and that a series of plausible hypotheses had been manipulated into solid facts, while a vast array of real facts on the opposite side had been completely ignored.'[35]

Some observers were less convinced that naming a victor was possible. 'On the whole it seemed to be a drawn battle,' noted the future professor of zoology at Cambridge (1866), Alfred Newton (1829–1907), 'for both sides stuck to their guns.'[36] Writing to Huxley's son, Frederic William Farrar, Bishop of Durham, also believed that the scientific issue remained largely deadlocked. Indeed, he was particularly adamant that if Huxley scored points against Wilberforce it was on the question of good manners, that the 'Bishop had forgotten to behave like a gentleman.'[37] On the question of manners, however, *The Press* clearly thought that Darwin's supporters were the 'aggressors' and that Huxley was 'facetious' while Wilberforce, on the other hand, 'brought all the well-known powers of his eloquence to substantiate the permanence of species.' *The Press* even justified Wilberforce's question of ancestry, arguing that the bishop 'had a right to the answer, that the commencement was not made by him.' However, much like Farrar and Newton, *The Press* was not about to crown a victor in this debate:

> In the midst of the discussions on this and other topics, involving at their root questions of the most vital, it has been pleasant to mark, for the most part, a spirit of wide and wise toleration, – as if truth and not any narrow party-victory were the earnest search of all. It has been, indeed, of special interest in the present meeting held at Oxford to see, as it were, the ancient city of the University open its arms in friendly embrace to the younger sons of science; to see, as it were, the greeting of ages mediaeval with times present; to mark how well it is possible for the Christian, the classic and the scientific to co-operate in the one grand end – the advancement of man and the glory of God.

According to *The Press*, the debate itself was not evidence of some grand conflict between science and religion – to the contrary, it was evidence of 'wide and wise toleration,' a spirit of cooperation between the 'Christian' and 'the scientific.'[38] Of course, that was the underlying belief of natural theology: that Christianity and science are complementary rather than conflictual. Even though *The Press* did not crown a victor, its coverage favoured a view of science that was at odds with the view promoted by Huxley and his friends.

Perhaps the more interesting question to ask about the debate is not 'Who won?' but rather 'What impact did the debate have on scientists?'

Henry Baker Tristram (1822–1911), the distinguished ornithologist, had been the first naturalist to publicly accept and apply Darwinian natural selection prior to the publication of the *Origin;* that had been on the basis of the joint Darwin–Wallace papers given at the Linnaean Society in July of 1858. In his article, 'On the ornithology of Northern Africa,' published in October 1859, he had illustrated Darwin and Wallace's theory of natural selection by discussing variations in the larks and chats of northern Africa. It should be noted, however, that Tristram was a deeply religious man and that, unlike Darwin, he was not vague about evolution being a product of God's handiwork; indeed, he argued that 'every peculiarity or difference in the living inhabitants of each country is admirably adapted by the wisdom of their beneficent Creator for the support and preservation of the species.'

Yet within a year of the *Origin*'s publication, Tristram had become an outspoken anti-Darwinian even though he had been the first to publish in support of natural selection. What had happened to reconvert Tristram? The simple answer: the Oxford debate.

Instead of supporting the Darwinians in the face of clerical attack, Tristram 'waxed exceedingly wroth' as the debate went on. As his friend Alfred Newton explained, this was not because of anything Wilberforce had said, but because of the response of the Darwinians. Tristram maintained that the followers of evolution could only support their doctrine through 'noise and sneers' directed at Wilberforce and 'everyone else who did not subscribe to the infallibility of the God Darwin and his prophet Huxley.' He now saw Darwinism as a simple 'renovation of Lamarck [and] one blind plunge into the gulph of atheism and the coarsest materialism.' The bishop's arguments clearly carried weight with Tristram, so much so that he was willing to denounce his own previous publication, which had supported Darwin's heretical theory.[39]

Tristram had been accompanied by his close friend Newton to the BAAS meeting of 1860. Newton had brought the joint Darwin–Wallace papers to Tristram's attention and had even summarized the way in which the theory could be brought to bear on Tristram's ornithological work, practically writing an abstract of what would become Tristram's aforementioned paper, 'On the ornithology of Northern Africa.' It is therefore likely that Tristram himself had not worked out the implications for his own religious beliefs before publishing on the matter; thus, when the bishop came out strongly against the theory, that caused him to reconsider his defence of Darwinism. Newton believed that Tristram's reconversion had been caused by 'a feeling of loyalty to the Bishop,' but it is possible that Tristram was simply not impressed with the Darwinians' responses.

Tristram would turn away from the Darwinian hypothesis after the Oxford debate; Newton, conversely, would become a hardened Darwinian despite his friend's reconversion. Before the Oxford debate, Newton had claimed that he was in a 'state of transition, [that] Darwin*oid* I might have remained for a whole geological aeon.' However, he wrote to Tristram, 'the Bishop's speech and article have caused me by a process of "natural selection" to become something better. I am developed into pure and unmitigated Darwinism.'[40] Unlike Tristram, Newton had been thoroughly convinced by Huxley and Hooker's defence of Darwin's theory, even while believing that the battle itself was a draw.

Debates rarely cause supporters of one side to see the merits of the other, much less cause anyone to change positions. The Oxford debate, however, caused at least one person to abandon a previous belief in evolution by natural selection while provoking someone in a state of transition to become a hardened Darwinian. By this score it would be difficult to choose a definite winner, if it could be put in such crude terms. Yet a particular narrative of the debate would become fairly common currency over the following years while the history of Darwinism was being written – a history that privileged one side's perspective over the other's while colouring the debate as a battle in a broader war between two great intellectual authorities, between old, irrational Christianity and younger, rational science. To understand the Oxford debate's place in the history of science, we must examine how it has been represented and remembered.

6 Remembering the Oxford Debate

> I hardly like to ask you to write, for I know how you are overworked; but I should rather like to hear a bit about the battle. I did not imagine that it would have turned up at Oxford; but I am now glad that I had no choice about going for I was utterly unfit. – The world surely will soon get weary of [the] subject and let us have some peace. Though, on the other hand, I do believe this row is the best thing for [the] subject.
>
> Charles Darwin to Thomas Henry Huxley, 3 July 1860

On the second level of the Natural History Museum in Oxford, a plaque next to an unassuming door states: 'A meeting of the British Association held 30 June 1860 within this door was the scene of the memorable debate on evolution between Samuel Wilberforce Bishop of Oxford and Thomas Henry Huxley.' Except for the plaque, however, there is no discernable evidence of how a great debate could have taken place behind that door. The once large space that held anywhere from seven hundred to one thousand excited audience members has been transformed into a long and narrow storage space, resembling a hallway more than a library; papers and folders and other miscellaneous items take up the space where 'the hottest of all debates' took place. For a historian looking to reconstruct the event known as the Oxford debate, it is a humbling experience to see that the room where it took place no longer exists, that the debate itself cannot be pictured where it happened, that it must be imagined by considering the sparse documents written at the time, documents that are themselves plagued with conflicting recollections and blatant exaggerations. Compounding the problem is that the debate was initially memorialized by one side, becoming in hindsight a key battle in the rhetorical war between science and religion, a Galileo Affair for the nineteenth century.[1] Early histories of the debate did little to deviate from the narrative set out by the supposed victors. Even now it is typical to find popular references to this debate in those mythologized terms.[2]

Immediately after the debate, Huxley, Hooker, and many of the other participants and observers took up Daubeny's invitation to visit his residence following the meeting. The day's events were of course discussed, and the 'sole topic was the battle of the "Origin."'[3] According to Isabel Sidgwick, 'I remember that some naïve person wished "it could come over again;" and Mr. Huxley, with the look on his face of the victor who feels the victory, put us aside saying, "Once in a life-time is enough, if not too much."'[4] Yet Huxley was more than happy to recollect the debate to willing listeners well after the day had come and gone. In September, a full two months after the debate, he wrote to his friend Frederick Dyster inquiring whether the 'rumours of Oxford' had reached him in Derby. 'It was great fun,' Huxley told him before detailing his encounter with 'saponaceous Samuel.' 'I think Samuel will think twice before he tries a *duel* with men of science again.'[5] Such stories spread.

The normally subdued Hooker was even more righteous than Huxley in his account of the debate and was less shy about using metaphors of a crusade-like nature. 'Sam Oxon got up and spouted for half an hour with inimitable spirit ugliness and emptyness and unfairness,' he wrote to Darwin. 'I saw that he was coached up by Owen and knew nothing ... He ridiculed you badly and Huxley savagely.' In Hooker's retelling, this was war. 'The battle waxed hot' he recounted. 'My blood boiled [and] I swore to myself that I would smite that Amalekite Sam hip and thigh if my heart jumped out of my mouth and I handed my name up to President (Henslow) as ready to throw down the gauntlet.' This was not subtle stuff. 'There I was cocked up with Sam at my right elbow, and there and then I smashed him amid rounds of applause – I hit him in the wind at the first shot in 10 words taken from his own ugly mouth.'[6] Most observers would recall that Hooker had appended a calm and scientific spirit to what had become an ugly affair during which well-regarded authorities traded low blows rather than useful commentary. In Hooker's recollection, however, he had been arm-and-arm with Huxley, helping pummel Soapy Sam and even offering up the knockout punch.

The young John Green, who was also present, recounted the debate in terms similar to both Huxley's and especially Hooker's. Green was clearly sympathetic to Darwin and evolution, and his account is flavoured accordingly. Green remembered that on being recruited to 'to hear the bishop smash Darwin,' he had responded: '"Smash Darwin! Smash the Pyramids," said I, in great wrath.' Remembering how 'Sammivel' 'got so uproarious as to pitch it into Darwin's friends,' Green declared that 'such rot never fell from episcopal lips before.' Huxley, in contrast, had been 'young, cool, quiet, sarcastic, scientific in fact and in treatment.' He 'gave his lordship such a smashing as he may meditate on with profit over his port at Cuddesdon.'[7]

No one seemed to get more pleasure out of the debate than Darwin himself. After he received word of the debate from Hooker, he wrote to Huxley to learn more about the encounter, and he wanted details:

> I had a letter from Oxford written by Hooker late on Sunday night, giving me some account of the awful battles which have raged about 'species' at Oxford. He tells me you fought nobly with Owen, (but I have heard no particulars) and that you answered the Bishop of Oxford capitally. Was Owen very blackguard? Hooker says that the Bishop turned me into ridicule and was very savage against you. – I hardly like to ask you to write, for I know how you are overworked; but I should rather like to hear a bit about the battle.[8]

Huxley's response has not been found, but Darwin was clearly delighted by it. Indeed, he read it twice and then sent it to his wife, knowing that she too would enjoy it. 'How durst you attack a live Bishop in that fashion?' Darwin wrote to Huxley from the safe confines of Sudbrook Park. 'I am quite ashamed of you! Have you no reverence for fine lawn sleeves? By Jove, you seem to have done it well.'[9] He was very grateful that he could rely on Huxley and Hooker to 'fight publicly' on his behalf, and because of this 'fully believe[d] that our cause will in the long run prevail.'[10]

Darwin also believed that the publicity generated by the debate would do wonders for the cause of evolution, and he was at the centre of an Oxford debate gossip network. He discussed the debate with Asa Gray (22 July 1860), Charles Lyell (30 July 1860), and the American geologist James Dwight Dana (30 July 1860).[11] Right after the debate he began circulating a story about Wilberforce that had been told to him by his eldest son William. Two men were discussing the encounter at Oxford, one of whom was Henry Fawcett, who happened to be blind. When Fawcett said in a loud voice that he did not believe Wilberforce had read a word of the *Origin,* the bishop overheard and was 'about to pitch into him, [but] saw that [Fawcett] was blind, and said nothing.'[12]

The debate became a common reference for Darwin and his inner circle, and they continued to allude to it for many years. When Fawcett published his account of the Oxford debate in *Macmillan's Magazine* at the end of 1860, Darwin made a point of circulating its contents, for it supported both Huxley's and Hooker's account of the event.[13] Huxley often mentioned his 'smashing' of the bishop long after the fact. In 1862, a full two years after the debate, he was still reminiscing about his victory, telling Darwin that someone in attendance at the debate told him 'it was worth while to come all the way to Oxford to hear the Bishop pummeled.'[14] That same year, Darwin wrote to Hooker reminiscing not about the debate itself but about Hooker's retelling of the debate and the joy he felt learning about his friends' great victory: 'To this day I remember keenly a letter you wrote to me from Oxford, when I was at the water-cure, and how it cheered me, when I was utterly weary of life.'[15] For Darwin and his friends, there was no doubt who won the debate, and there was both humour and hubris in their many references to it.

While there was little contention as to what had actually transpired at Oxford in June of 1860, Huxley more than Darwin and Hooker sought to correct any 'false' interpretations of the event. In January of 1861 he sent Wilberforce a copy of his recently published 'On the zoological relation of man with the lower animals' as 'full justification for the diametrical contradiction with which he heard Professor Huxley meet certain anatomical statements put forth at the first meeting of Section D during the late session of the British Association at Oxford.'[16] Huxley wanted to make sure Wilberforce understood he had been correct in his defence of Darwin's theory. He also wanted to control the narrative of the debate long after the fact.

When the time came to write Wilberforce's 'Life and Letters' following his death – the typical biographical format of Victorian intellectuals – the author, Wilberforce's son Reginald, found a brief entry in his father's diary describing his encounter with Huxley. He decided to summarize rather than quote its contents. 'The Bishop,' Reginald explained, 'made a long and eloquent speech condemning Mr. Darwin's theory as unphilosophical and as founded on fancy, and he denied that any one instance had been produced by Mr. Darwin which showed that the alleged change from one species to another had ever taken place.' The speech left a 'great impression,' Reginald contended.

He ended his brief summary with Huxley's retort to the Bishop's quip that 'he would not look at the monkeys in the Zoological as connected with his ancestors, a remark that drew from a certain learned professor the retort, "I would rather be descended from an ape than a bishop."'[17] Huxley was incensed, and he demanded that the sentence attributed to him be changed.

It is interesting that Huxley was not more angered by Reginald's interpretation of his father's comment regarding apes and ancestry, given that it was Wilberforce's indignant questioning of Huxley's ancestry that supposedly justified his retort – that in Lyell's words, 'the Bishop got no more than he deserved.'[18] Wilberforce's words were of course not recorded at the time and were clearly open to interpretation in later attempts to recount them accurately. Huxley himself remembered Wilberforce asking him whether he traced his ancestry through his grandfather; but other accounts suggest that Wilberforce asked him whether it was through his grandfather *or* his grandmother. It is also possible that Wilberforce was asking Huxley *when* in his family history the change from monkey to human occurred – at his mother, his grandmother, or further down the genealogical line. Whatever the case, most remember Wilberforce asking Huxley explicitly about his ape-like origins, and Reginald conveniently left that out of his narrative.[19] But Huxley was more angered by what Reginald claimed *he* had said than by the omission of the question from Wilberforce that sparked his response.

Huxley's response is just as difficult to determine. Many left the debate believing that Huxley had in fact said just what Reginald attributed to him. Huxley, however, immediately sought to correct such an interpretation, writing to Dyster that the reports being spread 'that I had said I would rather be an ape than a bishop' were false. He insisted that his response was much more convoluted:

> If then, said I, the question is put to me would I rather have a miserable ape for a grandfather or a man highly endowed by nature and possessed of great means and influence and yet who employs those faculties for the mere purpose of introducing ridicule into a grave scientific discussion – I unhesitatingly affirm my preference for the ape.[20]

John Green's report was quite similar, if slightly more eloquent:

> A man has no reason to be shamed of having an ape for a grandfather. If there were an ancestor whom I should feel sham in recalling, it would

> rather be a *man*, a man of restless and versatile intellect, who, not content with an equivocal success in his own sphere of activity, plunges into scientific questions with which he has no real acquaintance, only to obscure them by an aimless rhetoric, and distract the attention of his hearers from the real point at issue by eloquent digressions and skilled appeals to religious prejudice.[21]

Lyell admitted to hearing 'several different varying versions of [Huxley's] shindy' but said that it was to this effect:

> That if he had his choice of an ancestor, whether it should be an ape, or one who having received a scholastic education, should use his logic to mislead an untutored public, and should treat not with argument but with ridicule the facts and reasoning adduced in support of a grave and serious philosophical question, he would not hesitate for a moment to prefer the ape.[22]

There are many other versions of this retort available. Most of them certainly suggest *not* that Huxley preferred being related to an ape than to a bishop but rather that he preferred being related to an ape than to Wilberforce and perhaps to anyone else who might use their powers of eloquence to obscure the truth.[23] Witnesses' accounts have typically sided with Huxley regarding what exactly was said, which underscores that Huxley sought to control what was written about what he had said. Reginald Wilberforce, for his part, was forced to write an erratum for inclusion on the first page of the third volume of his *Life of Wilberforce*. It states that line 7 of page 451 ought to read as follows: 'If I had to choose between being descended from an ape or from a man who would use great powers of rhetoric to crush an argument, I should prefer the former.'[24] Whether or not Huxley actually won the debate is perhaps beside the point. He was certainly shaping the memory of the debate, and this mattered much more than anything he may or may not have said at Oxford.

Reginald Wilberforce would again clash with his father's foe about the events of Oxford. In 1887 the *Life and Letters of Charles Darwin* would appear, complete with a chapter written by Huxley reminiscing about the heady days of the 1850s and 1860s, when Darwinian evolution was being thrust onto the scene. Huxley, unsurprisingly, took shots at both Owen and Wilberforce via a critique of Wilberforce's *Quarterly Review* article on the *Origin*, in order to point out the nature of the opposition to Darwinian evolution. Reginald took offence at Huxley's attempt to blight the memory of his father, claiming in a letter to *The Times* that Huxley had in several instances misquoted his father. At issue, however, was not Huxley's distorted criticisms but why he felt the need to go on at length and fight a battle that was long over: 'What, it may fairly be asked, is the reason why these dead ashes are fanned into flame?' Ending there, Reginald would surely have made his point. Instead he went on to speculate about Huxley's reasons: 'Is it the memory of the debate in the Sheldon Theatre [sic] in 1860? Did the sharp lash of Bishop Wilberforce's eloquence sting so sharply that, though 27 years have passed, the recollection of the castigation then received is as fresh as ever?'[25] Clearly, Huxley wasn't the only one interested in controlling the memory of Oxford.

Huxley responded by rejecting Reginald's claims that he had somehow misread Wilberforce's review. He was unable to avoid commenting on the debate at Oxford, however. 'Those who were present at the famous meeting in Oxford,' he wrote, 'to which Mr. Wilberforce refers, will doubtless agree with him that an effectual castigation was received by somebody. But I have too much respect for filial piety, however indiscreet its manifestations, to trouble you with evidence as to who was the agent and who the patient in that operation.'[26] As usual, Huxley got in the last word while appearing to take the higher ground.

Of course, at this point, Huxley could be more subtle in his portrayal of the debate, given the narrative contained in the *Life and Letters of Charles Darwin* by Darwin's son Francis, which had been published earlier in the year (1887). In that work, Francis did little to describe the debate himself, largely letting the letters from participants and observers speak for themselves. He initially quotes from an anonymous 'eye-witness account of the scene,' an account that was certainly written by someone sympathetic to Huxley and Darwin and that parallels the one in Hooker's letter (quoted above) so closely in spirit and language that it was likely written by him: 'The Bishop … spoke for a full half-an-hour with inimitable spirit, emptiness and unfairness. It was evident from his handling of the subject that he had been "crammed" up to his throat, and that he knew nothing at first hand.' He then writes, as if plagiarizing his earlier letter to Darwin, that Wilberforce 'ridiculed Darwin badly, and Huxley savagely.' The winner of the debate could not be doubted: 'Huxley replied to the scientific argument of his opponent with force and eloquence, and to the personal allusion with a self-restraint, that gave dignity to his crushing rejoinder.'[27] In this account, it is interesting to note Huxley's 'self-restraint' – that his response was well within the confines of gentlemanly behaviour.

To flesh out the debate further, Francis quotes from John Green's letter, especially from Green's account of Huxley's response. Francis then returns to his anonymous eyewitness, who comments on Hooker's reply and on the break-up of the debate itself. No attempt is made in this passage to provide an alternative account of the debate, at least not from Wilberforce's perspective. This one-sided version of events, flavoured with the war cries of Green's letter and Hooker's anonymous eyewitness account, was the first attempt to historically reconstruct the debate, if one excludes the very brief summary made by Reginald Wilberforce a few years earlier. *The Life and Letters of Charles Darwin* was immensely popular, particularly for its genre, and the story of the Oxford debate, by that time long forgotten, was once again fodder for the educated public.[28] Reginald's letter to *The Times* was a clear attempt to provide an alternative account of a historical narrative that was being controlled by the 'victors.'

The Oxford debate was given a full twelve pages in Leonard Huxley's biography of his father, published in 1900. Leonard summarized the debate, touching on most of the familiar talking points: Huxley's earlier run-in with Owen; the irony of being cajoled by Robert Chambers to attend a debate he had been planning to skip; Draper's lecture; Wilberforce's diatribe; and Huxley's now legendary response. Unlike Francis Darwin's portrayal, however, Leonard's was buttressed by several eyewitness accounts. Alongside the typical extracts from Green, he provided accounts from Richard Greswell, Frederic William Farrar, and A.G. Vernon-Harcourt. He also included selections from other recently published narratives of the debate, such as the one by W.H. Freemantle as well as Isabel Sidgwick's anonymous article from *Macmillan's Magazine*.[29] Also included was an account written by Huxley himself in a letter to Francis Darwin in 1891. Leonard Huxley quoted this letter at length towards the end of his portrayal, and it added a sense of authority to the narrative as well as to the several excerpted first-hand accounts. 'I should say,' wrote the older Huxley, 'that Freemantle's account is substantially correct but that Green has the substance of my speech more accurately.'[30] From the grave, Huxley was able to guide the reader through the various accounts, pointing to the more accurate according to his own memory of the now thirty-year-old debate.

There is very little to distinguish the narrative thrust of Leonard Huxley's account from that of Francis Darwin's except that the former provides greater evidence to support the argument. Leonard would admit, however, that the day had not entirely been won; his father certainly turned a portion of the crowd to his favour, but at least half still sided with Wilberforce. Total victory was not what was important about the Oxford debate, however. What was important, Leonard Huxley argued, was 'the open resistance that was made to authority, at a moment when even a drawn battle was hardly less effectual than acknowledged victory. Instead of being crushed under ridicule, the new theories secured a hearing, all the wider, indeed, for the startling nature of their defense.'[31] Total victory would indeed come later.

In the months previous to the publication of the *Life and Letters of Thomas Henry Huxley*, Leonard received a letter challenging the typical portrayal of the debate as well as the one provided by the letters he had received. The letter was from Farrar, who would be one of the authorities cited in the *Life and Letters*, though his letter excerpted there focused on those who spoke prior to Wilberforce and was in some ways irrelevant to his main complaint. After seeing how the debate was being pieced together Farrar believed that most observers had gotten it all wrong. He contended that Wilberforce had been generally misquoted by Huxley's eyewitnesses. From the perspective of those eyewitnesses, Wilberforce's words 'did not appear vulgar, nor insolent, nor personal but flippant.' However, this was not actually the case. Wilberforce had erred not by questioning Huxley's ancestry generally, but rather by making the rather vulgar suggestion that Huxley would have to trace his ape ancestry not just through his grandfather's side but also through his grandmother's. 'The victory of your father,' then, argued Farrar, 'was not the ironical dexterity shown by him, but the fact that he had got a victory in respect of manners of good breeding. You must remember that the whole audience was made up of gentlefolk, who were not prepared to endorse anything vulgar.'[32]

Connecting Darwin's theory of evolution with the image of simian ancestry was a widespread strategy among anti-Darwinians because it challenged the respectability of evolution itself. As Gowan Dawson has shown, representations of ape-like humans in Victorian culture, while often humorous, were considered quite vulgar, and not simply because they appeared to debase humanity itself. This imagery implied that such a creature was not produced via evolution but via sexual relations between humans and apes. In 1873, *Punch* magazine provided a decidedly sexualized cartoon of the exchange between Huxley and Wilberforce. In what appears to be a smoky gentleman's club, two men are discussing human ancestry. The 'figurative party' exclaims that 'So long as *I* am a Man, Sorr, what does it matther to me whether me *Great-Grandfather* was an anthropoid ape or not, Sorr!' The 'Literal Party' responds in an aristocratic lisp: 'Haw! Wather disag-weeable for your *Gwate Gwandmother*, wasn't it!'[33] The implication, of course, is that the first man is related to an ape in a much more literal sense than evolution implies – that his grandmother had sexual relations with his ape ancestor. While it is unclear whether Wilberforce had this sort of sexualized imagery in mind when he asked Huxley through which grandparent he traced his simian ancestry, there was certainly something vulgar about his question. And as far as Farrar was concerned, Wilberforce's strategy of connecting evolution with recent simian ancestry did not have the desired effect. By questioning the respectability of evolution in such a manner, Wilberforce had, in the words of Farrar, 'forgotten to behave like a gentleman.'[34]

Farrar went on to say that it was not Huxley's retort to Wilberforce but rather Hooker's speech that 'made its mark *scientifically* on the meeting.' Yet that speech was mentioned only briefly in Leonard Huxley's narrative. Farrar continued by criticizing the reliance on the account of John Green, an account heavily relied upon by Francis Darwin as well: 'The Spiteful narrative of your quote from J. R. Green (the *historian*), is hardly worthy of him!'[35]

By this point, however, the history of the debate had already been written, and much of Farrar's letter would remain unpublished in the possession of Leonard Huxley until historians late in the twentieth century dug it up while visiting the Huxley archives at Imperial College, University of London. Until then, the general interpretation of the debate would remain in the form provided by both Leonard Huxley and Francis Darwin, as they were told it by their fathers and their fathers' friends. In that form, it had become a key victory in the rhetorical battle between science and religion.

By then, and throughout the 1870s, works had begun to appear describing the supposed battle between science and religion. The first of these to make its mark was written by none other than John William Draper, the forgotten participant in the Oxford debate, indeed, the man whose paper had provided the pretext for the debate.[36] In 1874 he published his *History of the Conflict between Science and Religion*, in which he set out to show the detrimental effects of religion – Catholicism in particular – on scientific progress. The debate between Huxley and Wilberforce was never mentioned in this work, which focused largely on the early modern period of Copernicus, Galileo, and Newton; yet it is clear that present times weighed heavily on Draper's interpretation. He had made it clear in his discussion at Oxford that he believed science needed to be purged entirely of any and all religious beliefs, and his 1874 book was an attempt to use historical examples as evidence of religion's anti-scientific nature. Most important, however, was Draper's delineation of the historical relationship between science and religion, a relationship he described as one of 'conflict.' In this way he nullified any other possible descriptions while reading into the past the conflict he had witnessed in his own life.

Twenty years later, Andrew Dickson White published his two-volume *History of the Warfare of Science with Theology in Christendom* (1896), a work that, even more explicitly than Draper's, projected the controversies of his own day into the past. This book begins with a history of the war between Creationists and evolutionists, describing how evolutionary thinking was slowly accepted despite the opposition put up by Creationists, who insisted that the Bible's account of species origin was authoritative. The chapter ends with a description of the Oxford debate as the last battle in a long war in which the evolutionists emerged victorious. Wilberforce 'disported himself,' argued White, in his attempt to outmatch the science of Darwin. Carried away by the 'tide of popular applause,' the bishop mistakenly 'congratulated himself ... that he was not descended from a monkey.' 'If I had to choose,' White quoted Huxley, 'I would prefer to be a descendant of a humble monkey rather than of a man who employs his knowledge and eloquence in misrepresenting those who are wearing out their lives in search of truth.' According to White, 'this shot reverberated through England, and indeed through other countries.'[37] Scientists would no longer genuflect in the presence of episcopal authority.

The Oxford debate did 'reverberate' throughout England and other countries, but not because Huxley put ecclesiastical authority in its place. In fact, it was a rather narrow debate among a few scholars that evolved into a theatrical narrative told in letters and in conversations, in books and in periodical articles, until it came to symbolize a shifting reality wherein evolution was becoming scientific orthodoxy, largely replacing theories that promoted species as individual divine creations. Indeed, Owen's archetype theory was quickly forgotten, with many of its original subscribers coming over to the evolutionary camp.

Even Owen, later in his life, began to be less vague about his adherence to evolution, though he would be careful to avoid committing himself to Darwinian evolution.

This shift towards the general acceptance of evolution can be seen quite clearly by comparing the science examination questions at Cambridge in the years surrounding the publication of the *Origin*. In 1851 a question asked: 'Reviewing the whole fossil evidence, shew that it does not lead to a theory of natural development through a natural transmutation of species.' By 1873 students were being asked to adopt a diametrically opposed position and to assume 'the truth of the hypothesis that the existing species of plants and animals have been devised by generation from others widely different.'[38] Well before that date, however, evolution had become orthodoxy at the annual meetings of the BAAS. Darwin's key mechanism of evolutionary change, natural selection, continued to be scrutinized, even by many of his closest supporters – including Huxley – but evolution itself tended to be taken as fact by scientists, at least by the time of the 1863 BAAS meeting. The meeting in Cambridge the previous year, it has been said, 'witnessed the last determined resistance of the anti-Darwinians and their ultimate defeat.'[39]

Despite their success in dismantling the previous compromise between science and religion, many scientific naturalists continued their drive to further secularize science. This campaign involved reminding the public of great battles endured in the past, battles that pitted a rational science against an irrational and powerful religion. In 1866 Hooker used his evening address at the BAAS meeting at Nottingham to describe the hostile environment that Darwin and his followers (Hooker referred to them as 'missionaries') had been compelled to endure in order to have their voices heard over the religious orthodoxy ('savages' with primitive beliefs), and he clearly had the meeting of 1860 in mind. 'The missionaries attempted to teach them, amongst other matters, the true theory of the moon's motions,' Hooker explained, drawing a not so subtle parallel between Darwin and Copernicus. 'The priests,' Hooker continued, 'attacked the new doctrine, and with fury, their temples were ornamented with symbols of the old creed, and their religious chants and rites were worded and arranged in accordance with it.' 'The medicine men' were divided on the issue, and several came to believe the new theory of the moon, and before long the new theory became 'accepted as fact, and the people applauded the new creed! Do you ask what tribes these were, and where their annual gatherings took place, and when? I will tell you. The first was in 1860, when the Derivative doctrine of species was first brought before the bar of a scientific assembly, and that the British Association at Oxford; and I need not tell those who heard *our* presiding Satchem's address last Wednesday evening that the last was at Nottingham.'[40] That Hooker's speech was met with uproarious applause rather than boos should go some way in suggesting the ascendance of Darwinian evolution within what had been an old boys' club dominated by natural theologists. Indeed, Darwin's theory dominated the meeting that year, with the Anglican *Guardian* reporting that it was 'impossible to pass from Section to Section without seeing how deeply [Darwin's] views have leavened the scientific mind of the day.'[41] Yet despite the place that had been won for Darwin's evolutionary theory, Hooker still felt the need to offer up recent history as a parable for the battle of good versus evil, science versus religion.

In 1874 the BAAS meeting was held in Belfast. There John Tyndall gave his Presidential Address, later known famously as the 'Belfast Address.' Tyndall, who was a close friend of Huxley, Hooker, and Darwin and who was an outspoken Darwinian, used his address as a pulpit to argue for the authority of science over religion. He gave a brief history of the relationship between the two and, relying on authorities such as Draper, described that relationship as one of long conflict. He suggested that the initial opposition to Darwinian evolution was religiously motivated and clouded by irrational beliefs – beliefs that contested those facts that challenged faith. That he was delivering his speech in Belfast made his argument all the more bold, given that city's image as a stronghold of Christianity. A close reading of the address reveals that Tyndall directed most of his ire towards Roman Catholicism; however, many listeners (and later readers) of his address believed that he was directing his comments at *all* religions, given his repeated use of the term 'theology.'[42] It should also be noted that two years later, Tyndall penned the prefatory note to Andrew White's *The Warfare of Science* (1876), an earlier version of what would become his *History of the Warfare of Science with Theology in Christendom*.

If it appears that there was a campaign to rid science of all theological remnants and to historicize the relationship between science and religion as a great battle, that is because there was in fact a campaign. The power and influence – and the sheer numbers – of young scientists expanded rapidly in the decades following publication of the *Origin*. A group of these young reformers began meeting on a regular basis, forming the X-Club in 1864. That group included Huxley, Hooker, Tyndall, Herbert Spencer, and other evolutionists, who together sought to extend their scientific influence and to promote their vision of a community devoted to values derived from rational science.[43] Many of the club members sought to extend the realm of science into what had generally been regarded as the religious sphere, in some ways bringing to pass Wilberforce's chief concern at Oxford – that Darwin's theory was essentially Darwin*ism*, a set of metaphysical beliefs in contradistinction to Christianity, masquerading as scientific fact. One result was that certain aspects of the debate acquired a distinctly political resonance. Indeed, from this perspective it is difficult not to perceive a certain ideological taint to Huxley and colleagues' uphill battle against a more powerful religious orthodoxy. Yet for the most part it is Huxley's version of events that has been passed down through the generations.

'In 1860,' notes Paul Johnson in *Offshore Islanders* (1972), 'when the Association for the Advancement of Science met in Oxford, Wilberforce, already known by the devastating name of Soapy Sam, was ill-advised enough to challenge Darwinism in open debate with T. H. Huxley, an encounter which his biographer wisely glosses over. The same year, he attacked a harmless volume of advanced theology, *Essays and Reviews*, in the pages of the *Quarterly* (for which he was paid the handsome sum of 100 guineas).'[44] These two sentences are filled with assumptions about the debate itself and the period surrounding Darwin's *Origin*. Johnson clearly took a one-sided view of events – events he has interpreted from a crude reading of the evidence.

The suggestion that *Essays and Reviews* was 'harmless' indicates an ignorance of the religious controversies surrounding its publication. If Protestantism is more affected by biblical criticism than Catholicism, as Johnson later suggests, then a book written by clergymen supporting contemporary historical and scientific findings that directly contradict scripture would in 1860 have been viewed from certain quarters as harmful. That Wilberforce's attack on Darwin's work was in some ways informed by the storm surrounding *Essays and Reviews* should go some way towards helping the present-day reader understand Soapy Sam's motives. The world he was born into and in which he grew up, and upon which he based his convictions and profession, was being turned upside down in the space of a few years. As well, his nickname was not always interpreted as 'devastating'; as we have seen, 'Soapy' began to take on a pejorative meaning only *after* the legend of the Oxford debate. Johnson also assumes that Wilberforce lost the Oxford debate – that he was 'ill-advised enough to challenge Darwinism in open debate.' As we have seen, it isn't so certain that he was defeated on that occasion. But presenting Wilberforce as the evil antagonist allows the historian to avoid coming to terms with Wilberforce's convictions and criticisms. Because of that presentation, however, this image of Wilberforce has endured.

History books, and other media, are full of portrayals that play loose in the same manner. For example, in 1978 the BBC aired a series titled *The Voyage of Charles Darwin*. One instalment, on 'Darwin's Bulldog,' included a dramatization of the Oxford debate derived entirely from Leonard Huxley's account, with no mention that the various sources contradicted one another. In that episode the debate comes to an end not with Hooker's reply but with that of Huxley, who is portrayed just as John Green described – as young, cool, and handsome. Wilberforce is dressed as the vaudeville villain – dark, leering, and ugly. There is no suggestion that Wilberforce might have come away with a partial victory. Indeed, he mutters to himself following Huxley's triumphant retort: 'Curses, foiled again.'[45]

Similarly, after Darwin's house in Downe was turned into a tourist attraction by English Heritage in 1996, a second-floor room called 'Controversy' was devoted to the many controversies engendered by both the *Origin* and *Descent of Man*. A display is devoted to describing the Oxford debate in unproblematic terms, complete with life-size cartoon cut-outs of Wilberforce and Huxley from *Vanity Fair*. In large, bold letters is Wilberforce's question regarding ancestry, followed by Huxley's retort. Most interesting, however, is that the debate itself is heard over the speakers in the room. At first one seems to be listening to a boring lecture being given to a restless audience. It must be that of Draper, but his 'nasal Yankeeism' is barely audible as the crowd grows more and more impatient. Someone can be heard shouting 'Get on with it man!' Draper's voice again becomes audible, but he has somehow morphed into Mr Dingle: 'Let this point A be man, and let that point B be the mawnkey.' Draper/Dingle is shouted down by calls of 'mawnkey!' and is never quite allowed to finish his speech. At this point someone shouts 'Make way for the good Bishop of Oxford!' as if Wilberforce has just shown up and halted all proceedings in order to give his sermon. 'They say Soapy Sam is out for Huxley's blood!' a voice exclaims. 'Mr. Darwin's casual theory is actually a shameful, unsubstantiated contestation of the Divine Revelation of the Bible,'

Wilberforce protests. Someone can be overheard snickering that he doubts that Wilberforce has even read the book – Hooker's eloquent defence of evolution has become a mere scoff from the audience. Wilberforce ends his speech by noticing that 'Professor Huxley is with us. You sir are a staunch advocate of Mr. Darwin's imaginings. Is it through your grandmother or grandfather that you claim descent from an ape?' After Wilberforce is chastised by Henslow, the meeting's chair, for 'using [the discussion] as a pulpit for puritan evangelicalism,' Huxley responds that he 'would rather be related to a miserable ape than a man who prostituted the gifts of culture and eloquence to the service of prejudice.' Immense cheers follow and then fade as the tape loops back to Draper's boring lecture.

No mention is made that the recording is a dramatization, that the words spoken by Huxley and Wilberforce are mere approximations to what they might have said, that the debate (seventeen years before Edison's first recording) was not actually recorded. For that moment, the visitor to Down House has the feeling of being transported back to Victorian Britain, back to the heady days of debate and battle, to the days when science and religion squared off in a violent war. But of course we are not actually hearing the debate as it actually happened; rather, we are being offered a mythologized version as it was remembered by selective participants, transformed into legend in succeeding years, and codified into history following Darwin's 'Life and Letters' and those of Huxley.[46]

About the same time that the BBC was promoting a crude and one-sided version of events for television, a handful of historians of science began to examine the Oxford debate and its rendition in the popular imagination with a more sceptical eye. Documents long since ignored were examined, such as newspaper reports of the BAAS meeting. Those reports, when they discussed the debate at all, did so in very different terms and often supported a very different narrative of the debate than what was found in the letters – letters on which historians had long relied. The narrative presented in the previous chapter relies on both the letters and the newspaper reports. As shown there, some of the reports seemed to suggest that Wilberforce had the better of the debate. More significantly, at least one report suggested that the debate was symbolic of the harmonic relationship between science and religion. Another presented the debate as a jovial rather than a war-like affair.

Taken together, the various reports and letters offer a picture very different from the one that has long been presented to the general public. For comparison purposes, here is a brief summary of the typical narrative of the debate as presented throughout the nineteenth century and most of the twentieth:

1 Wilberforce argued that the *Origin* contradicted scripture and ridiculed Huxley by questioning his ancestry – a clever tactic to sway an audience unlikely to support humankind's evolution from apes.
2 Huxley defended the scientific merits of evolution and humorously exposed Wilberforce's use of Christianity to obscure the truth.
3 The audience roared in approval of Huxley's defence and was largely swayed to an evolutionary view of species. The Darwinians clearly won the day.
4 The debate was a crucial episode in the battle fought by evolutionists against a powerful and unscientific established Church, pitting scientists against clerics.

By considering all the available evidence, we should be able to change the previous assumptions about the debate to the following:

1 Wilberforce challenged the methodology of the *Origin* and charged that the text was unphilosophical. He then questioned Huxley's ancestry, following up on a statement made by Huxley in a debate with Owen a few days earlier.
2 Huxley suggested that he would rather be related to an ape than to a man who would obscure the truth. There is evidence that this response was not heard by many in the crowd, that he was unable to throw his voice over such a large and loud assembly.
3 Huxley and Wilberforce were not the only speakers. Several others spoke against and in favour of evolution, and it is most likely that it was Hooker's speech, rather than Huxley's, that left its mark on the crowd.
4 This battle was likely a draw rather than an outright victory for either party. Also, the battle was not necessarily between clerics and scientists but between generations: a younger generation supporting Darwin and younger scientists, and an older generation supporting Wilberforce and conservative scientists.

Extracting the debate's significance from all of this is much more difficult than it used to be. No longer can it be cast in terms of science versus religion. But we can say that its impact to this day lies in how it was remembered rather than in what actually happened. From that perspective, the debate has the following points of significance:

1 The debate was cast in a dramatic format of heroes and villains, of good and evil, easily fitting within the binary narrative of science versus religion that developed in the second half of the nineteenth century. Indeed, the Oxford debate became a touchstone of this narrative.
2 Stories of the clash affected the relationship between science and religion in a negative way. It placed a strain on the argument that science and religion could coexist.
3 The debate generated great interest in Darwinism and emboldened the Darwinists. Indeed, Darwin was certain that the debate would serve his theory well, and he was right.[47]

Clearly, the debate's significance is much more complicated to explain when we consider not just what happened but how the debate itself was remembered. It should be cast not as an example of a battle in the war between science and religion but rather as a story used as a tool to support the professionalizing endeavours of a younger generation of scientists who saw religion as an impediment to their goals. The debate itself is no longer as significant as the stories told about it and how it has been remembered.

For professional historians of science, the Oxford debate is just one of many examples in the history of science and Christianity of an event that is far too easily fictionalized as a dramatic tale complete with heroes and villains, with theatrical battles, and with overblown consequences that supposedly changed the world.[48] The popular imagination seems to envision a grand battle between science and religion that has raged throughout history; and as we have seen, the Oxford debate fits quite easily within that sort of grand narrative. Because of this, professional historians of science have tended to minimize the debate's significance in an attempt to balance the scale, as it were, turning an event supporting the great battle narrative of science versus religion into a debate that was significant of very little or nothing at all.[49] James Secord has called the debate 'a minor incident, later raised to mythical dimensions by the need to make the Darwinian debate look more heated than it actually was.'[50] But isn't that precisely what makes the debate significant?

It is impossible to divorce the cultural memory of the debate from what actually happened in the Natural History Museum at Oxford University on 30 June 1860. The debate will continue to be remembered as a symbol of the controversy surrounding Darwinian evolution, and so it should. And as the terms of this controversy change, so too will the cultural memory of the debate itself. What I have tried to do is direct attention towards the personal and professional aspects of this controversy by examining the perspectives of each of the main participants in the Oxford debate. There is no doubt that the Oxford debate was symbolic of conflicts both within and between science and religion; but it was also – what is perhaps more important – about personal jealousies and professional rivalries, politics and power, tactics and strategies. Perhaps this tells us something about how science gets done as well.

Regarding Wilberforce, the debate must be understood in the context of a controversy within the Anglican Church between conservative and liberal factions – in other words, in terms of a conflict within religion itself. Darwinism, Soapy Sam believed, was aligned with Broad Churchmen who were too willing to cede religious ground to scientific authorities. Furthermore, when we consider Wilberforce's family history as well as the way in which he sought to devote himself to the Church of England in response to his wife's death, we arrive at a much fuller and more sympathetic account of his opposition to Darwinian evolution.

Richard Owen's opposition to Darwin's hypothesis must be understood as reflecting the powerful patronage system to which he was wedded. Owen had to be much more careful than someone like Darwin when interpreting his evidence, given that his funding was largely contingent on pleasing a scientific orthodoxy. This was what Huxley and other young scientists so hated about the system of scientific knowledge in mid-Victorian Britain. It is also likely that Owen's violent tone in opposition to evolution had much to do with his rivalry with Huxley.

Hooker's friendship with Darwin, it has been suggested, was of utmost importance in his defence of his friend's theory. Indeed, he was Darwin's closest confidant and in many ways helped Darwin develop the theory, by providing evidence for it even while challenging it during its conception.

Despite their different socio-economic positions, Huxley, Hooker, and Darwin shared scientific experiences, and all three were professional scientists. Huxley and to a lesser extent Hooker saw in the *Origin* a key scientific text that complemented their professionalizing agenda, one that extended the scope of rational science while excluding its religious dimension – a dimension that had long buttressed and supported the power and knowledge claims of the older generation of naturalists. Huxley acted at times as if he was on a crusade, a 'New Reformation' that would, in the words of Bernard Lightman, utilize the *Origin* as a 'Trojan horse within the fortress of science [and open] it up to the forces of unbelief.'[51] 'Theology and Parsondom,' Huxley wrote to Frederick Dyster a few months after the *Origin* was published, 'are in my mind the natural and irreconcilable enemies of Science. Few see it but I believe that we are on the Eve of a new Reformation and if I have a wish to live 30 yrs, it is to see the God of Science on the necks of her enemies.'[52] Is it any wonder that Wilberforce perceived in Darwinian evolution a creed, a metaphysical system of beliefs masked by the so-called purity of science?

Epilogue: The History of the Present

As suggested in the last chapter, the memory of the Oxford debate has been shaped by contemporary circumstances. It is fitting to end this book by discussing the current cultural climate that haunts any attempt to write the history of the Oxford debate. Historians are trained to suppress any personal biases in favour of the disinterested objectives of the historical community; however, as historians of historiography and science can attest, the possibility of achieving true objectivity remains a dream, even if it is a noble one.[1] It is better to recognize the ways in which the present weighs on the past instead of pretending that the historian somehow exists outside of history – outside of time and space – when he or she attempts to tell us what actually happened. Unfortunately, the transcendent authority of the historian is lessened when an admission of this kind is made, which is why historians, much like scientists, tend to ignore rather than confront the cultural contexts of their knowledge claims.

This project was in some ways drawn from an interest in present circumstances. In several jurisdictions in the United States in the early 2000s, a coalition of religious groups put pressure on legislators to allow schools to teach Creationism as a rival scientific theory alongside evolution. High-profile court cases followed, and the scientific merits of both theories were put on trial.[2] Could the Oxford debate, the first full-out public battle between Darwinians and Creationists, perhaps shed light on the present-day controversy? It did and it didn't.

Obviously, the position of Creationists and evolutionists has been greatly transformed. Evolutionary theory has only been strengthened in the 150 years since the publication of the *Origin,* thanks to the discovery of Gregor Mendel's work on heredity and the great strides made in genetics throughout the twentieth century, and this despite the ideological baggage that evolution has only recently begun to discard. Darwinian evolution is as stable today as Newtonian physics was in Darwin's day. Certainly we cannot ignore the possibility that it may one day be displaced by a better theory, but it would certainly seem odd to backtrack and teach theories that no longer make sense of the evidence, unless of course one is teaching history. Creationism, in other words, is no longer considered a plausible scientific theory within the international scientific community. Creationists themselves realize that they cannot go back to the days prior to the Scopes Monkey Trial (1925) and have the teaching of evolution removed from schools. Indeed, the best they can hope for is that Creationism will be taught as an alternative theory to evolution. It would be difficult to envision a battle between a present-day Huxley and Wilberforce, as they would never even be at the same debates, much less speaking in language the other side could understand.

This divide is only too clear in recent 'New Atheist' writings such as Richard Dawkins's *The God Delusion* (2006) and Christopher Hitchens's *God Is Not Great* (2007), two works that lambast Christianity as well as other religious systems while presenting science – Darwinian evolution in particular – as utterly incompatible with religious beliefs. These works, of course, speak only to the converted and actually share a great deal with histories of the late nineteenth century that sought to promote the great conflict between science and religion. They both ignore the fact that Darwin developed his theory within the framework of a scientific practice that was still deeply Christian. Many gentlemanly naturalists saw in evolution a theory that would inevitably undermine traditional Christian doctrine; yet others – for example, the Reverend Charles Kingsley – believed that Darwin's theory *strengthened* the power of God: 'We know of old that God was so wise that he could make all things; but, behold, he is so much wiser than even that, that he can make all things make themselves.'[3] Just as many naturalists in Darwin's day succeeded in reconciling their religious beliefs with a belief in evolution, so too have scientists in the present. 'Either half my colleagues are enormously stupid,' argued the late historian and evolutionary biologist Stephen Jay Gould, 'or else the science of Darwinism is fully compatible with conventional religious beliefs – and equally compatible with atheism.'[4] Even Huxley believed that there was no real conflict between science and religion when the two entities were properly understood; but he also contended that it was theology that was the problem, and he certainly avoided associating himself with atheism, given its association with working-class radicalism.[5] With that said, Huxley was interested not in bringing about a rapprochement between science and Christianity but rather in establishing concrete boundaries between the two. Part of this process involved creating a narrative where good science appeared at odds with religion, where powerful voices backed by Church authority appeared to be silencing disinterested considerations of natural history. As his review of the *Origin* in the *Westminster Review* suggested, all 'men of science' supported Darwinian evolution while all critics embraced the theory of special creation because they were deluded by 'Hebrew cosmogony.'[6] This narrative of a binary conflict is an attractive but overly simplistic one; it amounts to a myth that has long been appropriated to serve contemporary purposes, whether by Huxley to help undermine the cultural authority of Christianity in the Victorian era or by the New Atheists in their attempts to (re)establish the boundaries separating church and state in the post–9/11 United States.

Creationists, meanwhile, claim to be making a new argument by suggesting that evidence of design is evidence of a designer. There is, of course, nothing new about this argument. Michael Behe's bacterial flagellum has merely replaced William Paley's eye as the chief example of a perfectly designed organic system – a system so 'irreducibly complex' that it could only have been created through an abrupt act of creation. This is clearly the old logic of natural theology vaguely concealed by the apparently new scientific language of intelligent design. But rather than offering a new scientific alternative to evolution, Creationists offer – in Ian Hacking's words – a degenerating science, a research program that instead of 'reacting to counterexamples and difficulties by producing new theories that overcome the old hurdles,' merely avoids the problems by repeating old mantras. 'Degenerate programs' like Creationism, argues Hacking, 'paint themselves into smaller and smaller corners, skirting problems they'd prefer not to face. They seldom or never have a new, positive explanation of anything. In short, they teach us nothing.'[7] To suggest that such degenerating science be taught as a rival theory to evolution is ludicrous to say the least. But one must worry along with Hacking that present-day supporters of evolution who denounce Creationists out of hand while ignoring the past and present problems with evolutionary theory are forgetting what made Darwinian evolution such a progressive research program in the first place. From this perspective, the current evolution–creation struggle does seem to share a great deal with the Oxford debate; even though the players have changed and the two sides have grown further apart, it seems that the tactics have remained largely the same.

Analysing the heated rhetoric of either side, however, offers little insight into the nature of the debate itself. Merely considering the various arguments will not be enough if we want to *understand* the debate. To understand, we must consider a broader context when considering questions of an intellectual nature; and we must also consider the personalities, the struggles, the jealousies. Only then can we come to terms with the present debate surrounding Creationists and evolutionists, and perhaps only then can we consider the relationship between 1859 and today.[8] Yet there is no doubt that somewhere, Darwin's stomach still hurts.

Notes

Introduction

1 It should be noted that Darwin himself was ambiguous as to how the first living beings came into existence, whether by 'natural' or 'supernatural' causes.
2 Leslie Stephen, 'An Attempted Philosophy of History,' *Fortnightly Review* 27 (1880): 672–95.
3 See Bernard Lightman, '*Robert Elsmere* and the Agnostic Crisis of Faith,' in *Victorian Faith in Crisis: Essays on Continuity and Change in Nineteenth-Century Religious Belief*, ed. Richard J. Helmstadter and Bernard Lightman (London: Macmillan, 1990), 283–314.
4 *The Guardian*, 4 July 1860, p. 593, reported that Huxley and his ilk made it clear that 'they would rather be descended from apes than Bishops.'
5 Rivka Feldhay, *Galileo and the Church: Political Inquisition or Critical Dialogue?* (Cambridge: Cambridge University Press, 1995); and Mario Biagioli, *Galileo Courtier: The Practice of Science in the Culture of Absolutism* (Chicago: University of Chicago Press, 1993).
6 Edward J. Larson, *Summer for the Gods: The Scopes Trial and America's Continuing Debate over Science and Religion* (New York: Basic, 1997).
7 Christopher Hitchens, 'Equal Time: No Tax-Exempt Status for Churches that Refuse to Distribute Pro-Evolution Propaganda!' *Slate*, 23 August 2005. Kitcher's analysis of the debate in *Abusing Science: The Case against Creationism* (Cambridge: Cambridge University Press, 1982) is no better.
8 Edward Bulwer, *England and the English* (1833), vol. 2, p. 165; See also David Newsome, *The Victorian World Picture: Perceptions and Introspections in an Age of Change* (New Brunswick: Rutgers University Press, 1997), 1.

9 Charles Kingsley and John Morley quoted in Walter E. Houghton, *The Victorian Frame of Mind, 1830–1870* (New Haven: Yale University Press, 1957), 66–7.
10 Karl Marx and Friedrich Engels, *The Communist Manifesto* (New York: Penguin, 1967), 83.

1 Charles Darwin: Historian of Natural History

1 See Morse Peckham, ed., *On the Origin of Species by Charles Darwin: A Variorum Text* (Philadelphia: University of Pennsylvania Press, 1959), introduction and appendixes.
2 See Jack B. Morrell and Arnold W. Thackray, *Gentlemen of Science: Early Years of the British Association for the Advancement of Science* (Oxford: Oxford University Press, 1981).
3 On Darwin's sickness see Janet Browne, '"I Could Have Retched All Night": Charles Darwin and His Body,' in *Science Incarnate: Historical Embodiments of Natural Knowledge*, ed. Christopher Lawrence and Steven Shapin (Chicago: University of Chicago Press, 1998), 240–87; and Ralph Colp, Jr, *To Be an Invalid: The Illness of Charles Darwin* (Chicago: University of Chicago Press, 1977).
4 Adrian Desmond and James Moore, *Darwin: The Life of a Tormented Evolutionist* (New York: Norton, 1994).
5 Note that while Darwin lived in the village of 'Downe,' he referred to his dwelling as 'Down' House without the old English 'e.'
6 Charles Darwin, *On the Origin of Species*, ed. Joseph Carroll (1859; Orchard Park: Broadview, 2003), 95.
7 On Darwin's strategies of self-presentation see George Levine, *Darwin and the Novelists: Patterns of Science in Victorian Fiction* (Chicago: University of Chicago Press, 1991), 87–9.
8 Darwin and Wallace's initial correspondences have not been found, though there is reference to their early relationship in Charles Darwin to W.B. Tegetmeier, 29 November 1856, in *The Correspondence of Charles Darwin*, vol. 6, *1856–1857*, ed. Frederick Burkhardt and Sydney Smith (Cambridge: Cambridge University Press, 1990), 290.
9 Darwin to Charles Lyell, 18 June 1858, in *The Correspondence of Charles Darwin*, vol. 7, *1858–1859*, ed. Frederick Burkhardt and Sydney Smith (Cambridge: Cambridge University Press, 1991), 107.
10 Michael Ruse, *The Darwinian Revolution: Science Red in Tooth and Claw*, 2nd ed. (Chicago: University of Chicago Press, 1999), 188.

11 In his will, Francis Henry Egerton, eighth earl of Bridgewater, directed that £8,000 be paid to the persons who would undertake the task of writing eight volumes 'On the Power, Wisdom, and Goodness of God, as manifested in Creation,' to be nominated by the president of the Royal Society. Egerton died in 1829.
12 See, for instance, Jonathan R. Topham, 'Beyond the "Common Context": The Production and Reading of the Bridgewater Treatises,' *Isis* 89 (1998): 233–62; and John Hedley Brooke, 'Natural Theology and the Plurality of Worlds: Observations on the Brewster–Whewell Debate,' *Annals of Science* 34 (1977): 221–86.
13 Darwin, *On the Origin of Species*, 89.
14 Darwin, *The Origin of Species*, 6th ed. (1872; New York: Mentor, 1958).
15 Darwin, *On the Origin of Species*, 398.
16 Darwin, *The Origin of Species*, 6th ed., 459, emphasis added.
17 Darwin, *Autobiography*, ed. Nora Barlow (New York: Norton, 1969), 59.
18 See Ralph Colp, Jr, 'The Relationship of Charles Darwin to the Ideas of his Grandfather, Dr Erasmus Darwin,' *Biography* 9 (Winter 1986): 1–24.
19 Percy Bysshe Shelley, 'Preface,' in Mary Wollstonecraft Shelley, *Frankenstein; or, The Modern Prometheus*, ed. D.L. Macdonald and Kathleen Scherf, 2nd ed. (Peterborough: Broadview, 1999), 47.
20 [Richard Owen], 'Darwin *On the Origin of Species*,' *Edinburgh Review* 226 (April 1860): 487–532.
21 Darwin to Baden Powell, 18 January 1860, in *The Correspondence of Charles Darwin*, vol. 8, *1860*, ed. Frederick Burkhardt, Duncan M. Porter, Janet Browne, and Marsha Richmond (Cambridge: Cambridge University Press, 1993), 39.
22 Darwin to Baden Powell, 18 January 1860, in *Correspondence of Charles Darwin*, 8:40.
23 Curtis Johnson argues that these letters to Powell are evidence that the historical sketch Darwin would publish in 1860 was already largely written earlier in the 1850s and only rediscovered by Darwin after the first edition of the *Origin* was published. See Curtis N. Johnson, 'The Preface to Darwin's *Origin of Species*: The Curious History of the "Historical Sketch,"' *Journal of the History of Biology* 40 (2007): 529–56.
24 Darwin, *On the Origin of Species*, 397.
25 Darwin to Charles Lyell, 5 July 1856, in *Correspondence of Charles Darwin*, 6:167.

26 Darwin, *The Origin of Species*, 6th ed., 17–18.
27 Ibid.
28 Adrian Desmond, *The Politics of Evolution: Morphology, Medicine, and Reform in Radical London* (Chicago: University of Chicago Press, 1989).
29 Darwin, *The Origin of Species*, 6th ed., 18.
30 Ibid., 19–20.
31 Ibid.
32 This is not to say that *Vestiges* was entirely Lamarckian. As one of the anonymous reviewers of this manuscript points out, Chambers attached much importance 'to the protracted gestation of an embryo in inducing change' – a decidedly un-Lamarckian mechanism for evolution.
33 James A. Secord, *Victorian Sensation: The Extraordinary Publication, Reception, and Secret Authorship of* Vestiges of the Natural History of Creation (Chicago: University of Chicago Press, 2000).
34 Ibid., 501–3.
35 Quoted in ibid., 407.
36 Ibid., 409.
37 Loren Eiseley, *Darwin's Century: Evolution and the Men Who Discovered It* (New York: Anchor, 1961), 138.
38 Darwin, *The Origin of Species*, 6th ed., 20–1.
39 [Owen], 'Darwin *On the Origin of Species*,' 494.
40 Nicolaas A. Rupke, *Richard Owen: Victorian Naturalist* (New Haven: Yale University Press, 1994), 240.
41 Darwin, *The Origin of Species*, 6th ed., 21–2.
42 Ibid., 22–5 at 25.
43 Ibid., 25.
44 Adam Sedgwick to Charles Darwin, 24 November 1859, in *Correspondence of Charles Darwin*, 7:396.
45 Darwin to Lyell, 29 November 1859, and Darwin to Lyell, 2 December 1859, in *Correspondence of Charles Darwin*, 7:406, 409.
46 J.S. Henslow to Joseph Hooker, 10 May 1860, in *Correspondence of Charles Darwin*, 8:200.
47 Darwin to George Rolleston, 6 June 1860, in *Correspondence of Charles Darwin*, 8:244–5.
48 Darwin to Hooker, 15 May 1860, in ibid., 210.
49 Darwin to Hooker, 30 May 1860, in ibid.
50 Darwin to Lyell, 6 June 1860, in ibid., 242.
51 Darwin to Hooker, 12 June 1860, in ibid., 251.
52 Darwin to Hooker, 26 June 1860, in ibid., 268.

2 The Struggles of Soapy Sam

1 Bernard Lightman, *The Origins of Agnosticism: Victorian Unbelief and the Limits of Knowledge* (Baltimore: Johns Hopkins University Press, 1987), 118.
2 Standish Meacham, *Lord Bishop: The Life of Samuel Wilberforce 1805–1873* (Cambridge, MA: Harvard University Press, 1970), 12.
3 William to Samuel Wilberforce, 23 March 1822, in R.G. Wilberforce and A.R. Ashwell, *Life of Samuel Wilberforce*, 3 vols. (London: John Murray, 1880–3), 1:18–19.
4 Ibid., 1:3.
5 William to Samuel Wilberforce, 12 October 1823, in ibid., 1:20–1.
6 Samuel Wilberforce to Charles Anderson, 23 January 1838, quoted in ibid., 1:113–4.
7 *Essays and Reviews* sold 22,000 copies in two years, which was greater than the first twenty years of sales massed by the *Origin*.
8 Josef L. Altholz, *Anatomy of a Controversy: The Debate over* Essays and Reviews, *1860–1864* (Aldershot: Scolar, 1994), 32.
9 Baden Powell, 'On the Study of the Evidences of Christianity,' in *Essays and Reviews: The 1860 Text and Its Reading*, ed. Victor Shea and William Whitla (Charlottesville: University of Virginia, 2000), 251, 252, 258; and quoted in Pietro Corsi, *Science and Religion: Baden Powell and the Anglican Debate, 1800–1860* (Cambridge: Cambridge University Press, 1988), 217.
10 See chapter 1 for a discussion of this last sentence of the *Origin*.
11 Frederick Temple, 'The Education of the World,' in *Essays and Reviews*, 137–80; and quoted in Altholz, *Anatomy of a Controversy*, 31.
12 [Samuel Wilberforce], Review of *Essays and Reviews*, *Quarterly Review* 109 (1861): 248–305.
13 Ibid., 249–50.
14 'To the Reader,' in *Essays and Reviews*, 135.
15 [Wilberforce], Review of *Essays and Reviews*, 250–2.
16 [Samuel Wilberforce], Review of *On the Origin of Species*, *Quarterly Review* 108 (1860): 225–64, quote on 264.
17 Darwin to Hooker, 20 July 1860, in *The Correspondence of Charles Darwin*, vol. 8, *1860*, ed. Frederick Burkhardt, Duncan M. Porter, Janet Browne, and Marsha Richmond (Cambridge: Cambridge University Press, 1993), 293.
18 John Hedley Brooke, 'The Wilberforce–Huxley Debate: Why Did It Happen?' *Science and Christian Belief* 13 (2001): 139.

19 Darwin, *On the Origin of Species*, ed. Joseph Carroll (1859; Orchard Park: Broadview, 2003), 146.
20 [Wilberforce], Review of *On the Origin of Species*, 235.
21 Ibid., 240.
22 Darwin to Hooker, 20 July 1860, in *Correspondence of Charles Darwin*, 8:293. See also Darwin to Huxley, 20 July 1860, in ibid., 8:294.
23 [Wilberforce], Review of *On the Origin of Species*, 256–7.
24 Ibid., 258.
25 Peter Nichols, *Evolution's Captain: The Dark Fate of the Man Who Sailed Charles Darwin around the World* (New York: HarperCollins, 2003).
26 Emma Darwin to Charles Darwin [1839], in *Emma Darwin: A Century of Family Letters 1792–1896*, ed. Henrietta Litchfield, 2 vols. (London: John Murray, 1915), 2:173–4.
27 Ibid., 2:174.
28 Ibid.
29 Ibid., 2:175.
30 Darwin, *On the Origin of Species*, 143.
31 See Randal Keynes, *Annie's Box: Charles Darwin, His Daughter, and Human Evolution* (London: Fourth Estate, 2001).
32 Adrian Desmond and James Moore, *Darwin: The Life of a Tormented Evolutionist* (New York: Norton, 1994), 387.
33 Meacham, *Lord Bishop*, 57.
34 Quoted in ibid., 59.
35 Quoted in ibid.
36 Quoted in ibid.
37 *Natural Selection Not Inconsistent with Natural Theology* was the subtle title to Asa Gray's collection of essays on the *Origin* published in 1861.

3 Thomas Henry Huxley and Richard Owen

1 Thomas Henry Huxley, 'On the Theory of the Vertebrate Skull,' *Proceedings of the Royal Society of London* 9 (1857–9): 381–457; and Adrian Desmond and James Moore, *Darwin: The Life of a Tormented Evolutionist* (New York: Norton, 1994), 465.
2 Michael Ruse, *The Darwinian Revolution: Science Red in Tooth and Claw*, 2nd ed. (Chicago: University of Chicago Press, 1999), 197.
3 Charles Darwin, *On the Origin of Species*, ed. Joseph Carroll (Orchard Park: Broadview, 2003), 397.
4 Adrian Desmond, *Huxley: From Devil's Disciple to Evolution's High Priest* (London: Penguin, 1998), 3–7.

5 Ibid., 10–13.
6 Ibid., 34–5.
7 T.H. Huxley quoted in Leonard Huxley, *Life and Letters of Thomas Henry Huxley*, 2 vols. (New York: D. Appleton and Company, 1900), 1:25, 26.
8 T.H. Huxley, 'On the Anatomy and the Affinities of the Family of the Medusae,' *Philosophical Transactions of the Royal Society of London* (1843): 413.
9 'Degradation of Naval Surgeons,' *Lancet* 1 (1847): 680; Desmond, *Huxley*, 53.
10 Desmond, *Huxley*, 152–4.
11 T.H. Huxley to Eliza Huxley, 21 November 1850, in Huxley, *Life and Letters of Thomas Henry Huxley*, 1:69.
12 T.H. Huxley quoted in ibid., 1:72.
13 T.H. Huxley quoted in ibid., 1:74.
14 Desmond, *Huxley*, 172.
15 On Huxley's savaging of *Vestiges* see James A. Secord, *Victorian Sensation: The Extraordinary Publication, Reception, and Secret Authorship of* Vestiges of the Natural History of Creation (Chicago: University of Chicago Press, 2000), 500–4.
16 Desmond, *Huxley*, 195–215.
17 Richard Owen, 'On the generation of the marsupial animals, with a description of the impregnated uterus of the kangaroo,' *Philosophical Transactions of the Royal Society of London* (1834): 333–64.
18 Nicolaas A. Rupke, *Richard Owen: Victorian Naturalist* (New Haven: Yale University Press, 1994), 47–60 at 52. On Owen's strategies for self-presentation see Secord, *Victorian Sensation*, 421–6; and Aileen Fyfe, 'Conscientious Workmen or Booksellers' Hacks? The Professional Identities of Science Writers in the Mid-Nineteenth Century,' *Isis* 96 (2005): 194.
19 Paul White, *Thomas Huxley: Making the 'Man of Science'* (Cambridge: Cambridge University Press, 2003), 42; and Desmond, *Huxley*, 156.
20 Huxley to W. Macleay, 9 November 1851, in Huxley, *Life and Letters of Thomas Henry Huxley*, 1:102.
21 Huxley to Eliza Huxley, 20 May 1851 in ibid., 1:103.
22 Rupke, *Richard Owen*, 204; see also Michael Ruse, *The Evolution–Creation Struggle* (Cambridge, MA: Harvard University Press, 2005), 45.
23 Desmond, *Huxley*, 174; and Desmond and Moore, *Darwin*, 433.
24 Quoted in Secord, *Victorian Sensation*, 422.
25 Ibid.

26 Quoted in Desmond, *Huxley*, 193.
27 Desmond, *Huxley*, 193–4; and Desmond and Moore, *Darwin*, 433.
28 Ruse, *The Darwinian Revolution*, 143; White, *Thomas Huxley*, 52; and Desmond and Moore, *Darwin*, 437.
29 See Rebecca Stott, *Darwin and the Barnacle: The Story of One Tiny Creature and History's Most Spectacular Scientific Breakthrough* (London: Faber and Faber, 2004); and Ruse, *The Darwinian Revolution*, 184–8.
30 See Charles Darwin to T.H. Huxley, 11 April 1853, in *The Correspondence of Charles Darwin*, vol. 5, *1851–1855*, ed. Frederick Burkhardt and Sydney Smith (Cambridge: Cambridge University Press, 1989), 130.
31 Stott, *Darwin and the Barnacle*, 223; and Rebecca Stott, 'Masculinities in Nineteenth-Century Science: Huxley, Darwin, Kingsley, and the Evolution of the Scientist,' *Studies in History and Philosophy of Biological and Biomedical Sciences* 35 (2004): 199.
32 Darwin to Huxley, 23 April 1853, in *Correspondence of Charles Darwin*, 5:133.
33 Thomas Henry Huxley, 'Science,' *Westminster Review* 61 (1854): 254–70, esp. 264–5.
34 Desmond and Moore, *Darwin*, 433.
35 White, *Thomas Huxley*, 8.
36 This important passage is quoted in full in Desmond, *Huxley*, 657–8n20.
37 Desmond and Moore, *Darwin*, 465.
38 Ibid., 451–3.
39 Ibid., 465.
40 Quoted in Desmond, *Huxley*, 241.
41 Ibid., 244.
42 T.H. Huxley to Darwin, 23 November 1860, in *The Correspondence of Charles Darwin*, vol. 7, *1858–1859*, ed. Frederick Burkhardt and Sydney Smith (Cambridge: Cambridge University Press, 1991), 390.
43 Huxley, *Life and Letters of Thomas Henry Huxley*, 1:189–90.
44 Janet Browne, *Charles Darwin*, vol. 2, *The Power of Place* (London: Pimlico, 2003), 100.
45 Desmond and Moore, *Darwin*, 480.
46 [T.H. Huxley], 'The Origin of Species,' *Westminster Review* 17 (1860): 541–70; also published as Thomas H. Huxley, *Darwiniana* (New York: D. Appleton and Company, 1897), 22–79.
47 Huxley, *Darwiniana*, 52–3.
48 Ibid., 58.
49 Ibid., 79.

50 Darwin to C.R. Lyell, 10 April 1860, in *Correspondence of Charles Darwin*, vol. 8, *1860*, ed. Frederick Burkhardt, Duncan M. Porter, Janet Browne, and Marsha Richmond (Cambridge: Cambridge University Press, 1993), 153.

51 Darwin to C. R. Lyell, 10 April 1860, in ibid., 8:153.

4 Joseph Dalton Hooker and the Early History of a Great Friendship

1 Joseph Hooker to Charles Darwin, 2 July 1860, in *The Correspondence of Charles Darwin*, vol. 8, *1860*, ed. Frederick Burkhardt, Duncan M. Porter, Janet Browne, and Marsha Richmond (Cambridge: Cambridge University Press, 1993), 271.

2 Hooker to Darwin, 2 July 1860, in ibid.

3 Hugh Falconer to Darwin, 9 July 1860, in ibid., 8:282.

4 Ralph Colp, Jr, '"Confessing a Murder": Darwin's First Revelations about Transmutation,' *Isis* 77 (1986): 8–32.

5 Hooker quoted in Leonard Huxley, *The Life and Letters of Sir Joseph Dalton Hooker*, 2 vols. (London: John Murray, 1918), 1:136.

6 Hooker quoted in Adrian Desmond and James Moore, *Darwin: The Life of a Tormented Evolutionist* (New York: Norton, 1994), 314.

7 Darwin to Hooker, 11 January 1844, in *The Correspondence of Charles Darwin*, vol. 3, *1844–1846*, ed. Frederick Burkhardt and Sydney Smith (Cambridge: Cambridge University Press, 1987), 1.

8 Hooker to Darwin, 29 January 1844, in ibid, 3:5; and Jim Endersby, *Imperial Nature: Joseph Hooker and the Practices of Victorian Science* (Chicago: University of Chicago Press, 2008), 47.

9 Hooker quoted in Huxley, *The Life and Letters of Sir Joseph Dalton Hooker*, 1:46.

10 Ibid., 1:37.

11 Endersby, *Imperial Nature*, 32, 43.

12 James Clark Ross and Hooker quoted in Huxley, *The Life and Letters of Sir Joseph Dalton Hooker*, 1:41; and Endersby, *Imperial Nature*, 33.

13 Endersby, *Imperial Nature*, 33.

14 Huxley, *The Life and Letters of Sir Joseph Dalton Hooker*, 1:45.

15 Darwin to William Jackson Hooker, 12 March 1843, in *The Correspondence of Charles Darwin*, vol. 2, *1837–1843*, ed. Frederick Burkhardt and Sydney Smith (Cambridge: Cambridge University Press, 1986), 242.

16 Colp, Jr., '"Confessing a Murder,"' 12.

17 Huxley, *The Life and Letters of Sir Joseph Dalton Hooker*, 1:486.

18 Hooker quoted in *The Life and Letters of Charles Darwin*, ed. Francis Darwin, 2 vols. (New York: D. Appleton and Company, 1887), 1:387–8.

19 Hooker quoted in ibid., 1:388.
20 Darwin to Hooker, 25 September 1853, in *The Correspondence of Charles Darwin*, vol. 5, *1851–1855*, ed. Frederick Burkhardt and Sydney Smith (Cambridge: Cambridge University Press, 1989), 155.
21 Duncan M. Porter, 'On the Road to the *Origin* with Darwin, Hooker, and Gray,' *Journal of the History of Biology* 26 (Spring 1993): 19.
22 Charles Darwin, *On the Origin of Species*, ed. Joseph Carroll (Orchard Park: Broadview, 2003), 96.
23 Darwin to Hooker, 20 October 1858, *The Correspondence of Charles Darwin*, vol. 7, *1858–1859*, ed. Frederick Burkhardt and Sydney Smith (Cambridge: Cambridge University Press, 1991), 174; and Huxley, *The Life and Letters of Sir Joseph Dalton Hooker*, 1:500.
24 Michael Ruse, *The Darwinian Revolution: Science Red in Tooth and Claw* (Chicago: University of Chicago Press, 1979), 139–40; see also Porter, 'On the Road to the *Origin*,' 13.
25 Endersby, *Imperial Nature*, 157.
26 John Beatty, 'Speaking of Species: Darwin's Strategy,' in *The Darwinian Heritage*, ed. David Kohn (Princeton: Princeton University Press, 1982), 277.
27 J.D. Hooker to A. Gray, 27 August 1857, quoted in Porter, 'On the Road to the *Origin*,' 27.
28 Darwin to Hooker, 31 December 1858, in *Correspondence of Charles Darwin*, 7:228; and quoted in Huxley, *The Life and Letters of Sir Joseph Dalton Hooker*, 1:501.
29 See the letters quoted in ibid., 1:515–19.
30 Hooker to Harvey, 26 May 1860, in ibid., 1:518.
31 Hooker to Harvey, June 1860, in ibid., 1:519–20.
32 Hooker to Darwin, 2 July 1860, in *Correspondence of Charles Darwin*, 8:271.

5 The Oxford Debate

1 'Section D.—Zoology and Botany, Including Physiology,' *The Athenaeum*, 7 July 1860, pp. 25–6.
2 John Green tells us that on the Saturday morning before Draper's talk, he joined company with Jenkins, who said that he was off to Section D 'to hear the Bishop of Oxford smash Darwin.' J.R. Green to W. Boyd Dawkins, 3 July 1860, in *Letters of John Richard Green*, ed. Leslie Stephen (London: Macmillan & Co., 1901), 44.
3 T.H. Huxley to Frederick Daniel Dyster, 9 September 1860, T.H. Huxley Papers, vol. XV, fols. 115–18, Imperial College, Huxley Archives.

4 Joseph Hooker to Charles Darwin, 2 July 1860, in *The Correspondence of Charles Darwin*, vol. 8, *1860*, ed. Frederick Burkhardt, Duncan M. Porter, Janet Browne, and Marsha Richmond (Cambridge: Cambridge University Press, 1993), 271.
5 Carla Yanni, *Nature's Museum: Victorian Science and the Architecture of Display* (London: Athlone, 1999), 63.
6 The previous paragraph is based on Robert Fox's description in 'The University Museum and Oxford Science, 1850–1880,' in *The History of the University of Oxford*, vol. 6, ed. M.G. Brock and M.C. Curthoys (Oxford: Clarendon, 1997), 657–8.
7 'Section D.—Zoology and Botany, Including Physiology,' *The Athenaeum*, 14 July 1860, p. 64.
8 Green to Dawkins, 3 July 1860, in *Letters of John Richard Green*, 44.
9 Hooker to Darwin, 2 July 1860, in *Correspondence of Charles Darwin*, 8:270.
10 Michael Ruse, *Monad to Man: The Concept of Progress in Evolutionary Biology* (Cambridge: Harvard University Press, 1996).
11 John William Draper, *History of the Intellectual Development of Europe*, 2 vols. (New York: Bell and Daldy, 1863).
12 [Isabel Sidgwick], 'A Grandmother's Tales,' *Macmillan's Magazine* 78 (1898): 433.
13 'Section D.—Zoology and Botany, Including Physiology,' 65.
14 Rev. Farrar quoted in Leonard Huxley, *Life and Letters of Thomas Henry Huxley*, 2 vols. (New York: D. Appleton and Company, 1900), 1:196.
15 Rev. W. Tuckwell, *Reminiscences of Oxford* (London: Cassell and Company, 1901), 51.
16 The previous quotations by Wilberforce are recorded in 'Section D.—Zoology and Botany, Including Physiology,' 65.
17 Wilberforce's precise wording is difficult to determine, as there are a variety of slightly different reports, though the general interpretation seems to be the same. See J. Vernon Jenson, 'Return to the Wilberforce–Huxley Debate,' *British Journal for the History of Science* 21 (1988): 167, for an analysis of the various reports.
18 So Huxley told Darwin's son fifty years later. T.H. Huxley to Francis Darwin, 27 June 1891, quoted in Huxley, *Life and Letters of Thomas Henry Huxley*, 1:202.
19 Ibid.
20 Huxley to Dyster, 9 September 1860, Huxley Papers.
21 Henry Fawcett, 'A Popular Exposition of Mr. Darwin on the Origin of Species,' *Macmillan's Magazine* 3 (December 1860): 88.
22 [Sidgwick], 'A Grandmother's Tales,' 434.

23 'Section D.—Zoology and Botany, Including Physiology,' 65.
24 Huxley to Dyster, 9 September 1860, Huxley Papers.
25 Hooker to Darwin, 2 July 1860, in *Correspondence of Charles Darwin*, 8:271.
26 Darwin to Huxley, 3 July 1860, in ibid., 8:277.
27 Green to Dawkins, 3 July 1860, in *Letters of John Richard Green*, 42.
28 [Sidgwick], 'A Grandmother's Tales,' 434.
29 Quoted in Huxley, *Life and Letters of Thomas Henry Huxley*, 1:200.
30 Ibid., 1:204.
31 *Evening Star*, 2 July 1860, p. 3.
32 Huxley, *Life and Letters of Thomas Henry Huxley*, 1:202.
33 Samuel Wilberforce to Sir Charles Anderson, 3 July 1860, MSS Wilberforce, d. 29, fols. 30–2, Bodleian Library, Oxford University.
34 Balfour Stewart to David Forbes, 4 July 1860, quoted in Stephen Jay Gould, *Bully for Brontosaurus: Reflections in Natural History* (New York: Norton, 1992), 389.
35 *John Bull*, 6 July 1860, p. 422.
36 Alfred Newton, 'Early Days of Darwinism,' *Macmillan's Magazine* 57 (February 1888), 249.
37 F.W. Farrar to Leonard Huxley, 12 July 1899, Huxley Papers, vol. XVI, fols. 13–19.
38 'Literature and Art,' *The Press*, 7 July 1860, p. 656.
39 Henry Baker Tristram to Alfred Newton, 31 July 1860, in A.F.R. Wollaston, *Life of Newton* (London: John Murray, 1921), 121–2; I. Bernard Cohen, 'Three Notes on the Reception of Darwin's Ideas on Natural Selection (Henry Baker Tristram, Alfred Newton, Samuel Wilberforce),' in *The Darwinian Heritage*, ed. David Kohn (Princeton: Princeton University Press, 1985), 598. This episode is also recounted by John Hedley Brooke, 'The Wilberforce–Huxley Debate: Why Did It Happen?' *Science and Christian Belief* 13 (2001): 128–9.
40 Alfred Newton to Henry Baker Tristram, 30 July 1860, in Wollaston, *Life of Newton*, 121.

6 Remembering the Oxford Debate

1 Frank A.J.L. James has provided a careful and thorough summary of all the known sources on the Oxford debate in 'An "Open Clash between Science and the Church"? Wilberforce, Huxley, and Hooker on Darwin at the British Association, Oxford, 1860,' in *Science and Beliefs: From Natural Philosophy to Natural Science, 1700-1900*, ed. David M. Knight and Matthew D. Eddy (Aldershot: Ashgate, 2005): 171–93.

2 See, for instance, Cyril Aydon, *Charles Darwin* (New York: Carroll and Graff, 2003), 214–26.
3 Quoted in *The Life and Letters of Charles Darwin*, ed. Francis Darwin, 2 vols. (New York: D. Appleton and Company, 1887), 2:116.
4 [Isabel Sidgwick], 'A Grandmother's Tales,' *Macmillan's Magazine* 78 (1898), 434.
5 T.H. Huxley to Frederick Daniel Dyster, 9 September 1860, T.H. Huxley Papers, vol. XV, fols. 115–18, Imperial College, Huxley Archives, emphasis added.
6 Joseph Hooker to Charles Darwin, 2 July 1860, in *The Correspondence of Charles Darwin*, vol. 8, *1860*, ed. Frederick Burkhardt, Duncan M. Porter, Janet Browne, and Marsha Richmond (Cambridge: Cambridge University Press, 1993), 271.
7 Green to W. Boyd Dawkins, 3 July 1860, in *Letters of John Richard Green*, ed. Leslie Stephen (New York: Macmillan and Company, 1901), 44–5.
8 Darwin to Huxley, 3 July 1860, in *Correspondence of Charles Darwin*, 8:277.
9 Darwin to Huxley, 5 July 1860, in ibid., 8:280.
10 Darwin to Hooker, 2 July 1860, in ibid., 8:272.
11 Ibid., 8:298, 306, 303.
12 Darwin to Huxley, 30 July 1860, in ibid., 8:305.
13 Henry Fawcett, 'A Popular Exposition of Mr. Darwin on the Origin of Species,' *Macmillan's Magazine* 3 (December 1860): 81–92; Darwin to Huxley, 5 December 1860, in *Correspondence of Charles Darwin*, 8:514; and Darwin to Hooker, 6 December 1860, in ibid., 8:516.
14 Huxley to Darwin, 20 January 1862, in *Correspondence of Charles Darwin*, vol. 10, *1862*, ed. Frederick Burkhardt, Joy Harvey, Duncan M. Porter, and Jonathan R. Topham (Cambridge: Cambridge University Press, 1997), 33.
15 Darwin to Hooker, 30 June 1862, in ibid., 10:283.
16 T.H. Huxley to Samuel Wilberforce, 3 January 1861, T.H. Huxley Papers, vol. 121/118, box 121, Imperial College, Huxley Archives.
17 R.G. Wilberforce, *Life of the Right Reverend Samuel Wilberforce*, vol. 2 (London: John Murray, 1881), 450–1.
18 Charles Lyell to Sir Charles Bunbury, 4 July 1860, in *Life, Letters, and Journals of Sir Charles Lyell*, ed. Mrs. Lyell (London: John Murray, 1881), 335.
19 The various interpretations of what may or may not have been said at the Oxford debate are wonderfully deconstructed by J. Vernon Jenson, 'Return to the Wilberforce–Huxley Debate,' *British Journal for the History of Science* 21 (1988): 161–79.

20 Huxley to Dyster, 9 September 1860, Huxley Papers.
21 Green to Dawkins, 3 July 1860, in *Letters of John Richard Green*, 45.
22 Charles Lyell to Sir Charles Bunbury, 4 July 1860, in *Life, Letters and Journals of Sir Charles Lyell*, 335.
23 Jenson, 'Return to the Wilberforce–Huxley Debate.'
24 R.G. Wilberforce, *Life of the Right Reverend Samuel Wilberforce*, vol. 3 (London: John Murray, 1882), 1.
25 R.G. Wilberforce, 'Professor Huxley and the Life and Letters of C. Darwin,' *The Times*, 29 November 1887, p. 10. Note that the debate took place in the library in the Museum of Natural History, not the Sheldon Theatre.
26 T.H. Huxley, 'Bishop Wilberforce and Professor Huxley,' *The Times*, 1 December 1887, p. 8.
27 Quoted in *The Life and Letters of Charles Darwin*, 2:113–16.
28 Frank James makes the case that the debate itself had little impact on the wider public until it was recounted in the *Life and Letters of Charles Darwin*. See James, 'An "Open Clash between Science and the Church"?,' 175.
29 W.H. Freemantle, *Charles Darwin, his Life Told* (1892), 238; and [Sidgwick], 'A Grandmother's Tales,' 425–35.
30 T.H. Huxley to Francis Darwin, 27 June 1891, in Leonard Huxley, *Life and Letters of Thomas Henry Huxley*, 2 vols. (New York: D. Appleton and Company, 1900), 2:201.
31 Huxley, *Life and Letters of Thomas Henry Huxley*, 2:204.
32 Frederic William Farrar to Leonard Huxley, 12 July 1899, T.H. Huxley Papers, vol. XVI, fols. 13–19, Imperial College, Huxley Archives.
33 G.M. [George Du Maurier], 'The Descent of Man,' *Punch* 64 (1873), 217. This cartoon is reproduced and analysed in Gowan Dawson, *Darwin, Literature, and Victorian Respectability* (Cambridge: Cambridge University Press, 2007), 58–9.
34 Farrar to Leonard Huxley, 12 July 1899, Huxley Papers.
35 Ibid.
36 David N. Livingstone also points this out in his 'Re-placing Darwinism and Christianity,' in *When Science and Christianity Meet*, ed. David C. Lindberg and Ronald L. Numbers (Chicago: University of Chicago Press, 2003), 183–202.
37 Andrew D. White, *History of the Warfare of Science with Theology in Christendom*, 2 vols. (1896; Buffalo: Prometheus, 1993), 2:70.
38 Michael Ruse, *The Darwinian Revolution: Science Red in Tooth and Claw*, 2nd ed. (Chicago: University of Chicago Press, 1999), xii.

39 A.F.R. Wollaston, *Life of Alfred Newton* (London: John Murray, 1921), 123.
40 Hooker quoted in Janet Browne, *Charles Darwin*, vol. 2, *The Power of Place* (London: Pimlico, 2003), 271–2.
41 Adrian Desmond and James Moore, *Darwin: The Life of a Tormented Evolutionist* (New York: Norton, 1994), 536.
42 Frank Turner, 'The Victorian Conflict between Science and Religion: A Professional Dimension,' *Isis* 69 (1978): 373–4.
43 On the X-Club see Ruth Barton, '"Huxley, Lubbock, and Half a Dozen Others": Professionals and Gentlemen in the Formation of the X Club,' *Isis* 89 (1998): 410–44; and Roy M. Macleod, 'The X-Club: A Social Network of Science in Late-Victorian England,' *Notes and Records of the Royal Society of London* 24 (1970): 305–22.
44 Paul Johnson, *The Offshore Islanders* (London: Weidenfeld and Nicolson, 1972), 305.
45 Jenson, 'Return to the Wilberforce–Huxley Debate,' 176. See also Sheridan Gilley and Ann Loades, 'Thomas Henry Huxley: The War between Science and Religion,' *The Journal of Religion* 61 (1981): 285; Sheridan Gilley, 'The Huxley–Wilberforce Debate: A Reconsideration,' in *Religion and Humanism: Studies in Church History*, Vol. XVII, ed. Keith Robins (Oxford: Blackwell, 1981), 325; and John R. Lucas, 'Wilberforce and Huxley: A Legendary Encounter,' *Historical Journal* 22 (1979): 313.
46 The recording has since 2008 been removed and the displays revised to reflect recent interpretations.
47 Jenson describes similar points of significance in his 'Return to the Wilberforce–Huxley Debate,' 177–9.
48 David Philip Miller, 'The "Sobel Effect": The Amazing Tale of How Multitudes of Popular Writers Pinched All the Best Stories in the History of Science and Became Rich and Famous While Historians Languished in Accustomed Poverty and Obscurity, and How This Transformed the World,' *Metascience* (2002): 185–200.
49 Frank James claims that he sought 'deliberately to minimise the importance' of the debate in 'An "Open Clash between Science and the Church"?' 187.
50 James A. Secord, *Victorian Sensation: The Extraordinary Publication, Reception, and Secret Authorship of* Vestiges of the Natural History of Creation (Chicago: University of Chicago Press, 2000), 513.
51 Bernard Lightman, 'Science and Unbelief,' conference on Science and Religion around the World, Green College, University of British Columbia, May 2007. See also Lightman, 'Interpreting Agnosticism as a Nonconformist Sect: T.H. Huxley's "New Reformation,"' in *Science and*

Dissent in England, 1688–1945, ed. David M. Knight and Trevor H. Levere (Aldershot: Ashgate, 2004), 197–214.

52 Huxley to Frederick Daniel Dyster, 30 January 1859, quoted in Michael Ruse, *The Evolution–Creation Struggle* (Cambridge, MA: Harvard University Press, 2005), 125–6.

Epilogue: The History of the Present

1 Peter Novick, *That Noble Dream: The 'Objectivity Question' and the American Historical Profession* (Cambridge: Cambridge University Press, 1988); and George Levine, *Dying to Know: Scientific Epistemology and Narrative in Victorian England* (Chicago: University of Chicago Press, 2002).

2 Matthew Chapman, 'God or Gorilla: A Darwin Descendant at the Dover Monkey Trial,' *Harper's Magazine* (February 2006): 54–63. This was later expanded and published as *40 Days and 40 Nights: Darwin, Intelligent Design, God, OxyContin, and Other Oddities on Trial in Pennsylvania* (New York: HarperCollins, 2007).

3 Charles Kingsley quoted in John Hedley Brooke, 'Natural Theology and the Plurality of Worlds: Observations on the Brewster–Whewell Debate,' *Annals of Science* 34 (1977), 224.

4 Stephen Jay Gould, 'Impeaching a Self-Appointed Judge,' *Scientific American* 267 (July 1992): 118–21.

5 See Huxley's essays in *Science and Hebrew Tradition* (London: Macmillan and Company, 1893), 160–1; and Bernard Lightman, *The Origins of Agnosticism: Victorian Unbelief and the Limits of Knowledge* (Baltimore: Johns Hopkins University Press, 1987), 131–2. I thank an anonymous reviewer for making this point.

6 T.H. Huxley, *Darwiniana* (New York: D. Appleton and Company, 1897), 22–3, 53–4; and Paul White, *Thomas Huxley: Making the 'Man of Science'* (Cambridge: Cambridge University Press, 2003), 51.

7 Ian Hacking, 'Root and Branch,' *The Nation*, 8 October 2007, p. 27.

8 See Ruse, *The Evolution–Creation Struggle*, for a history of the debate throughout the last two centuries.

Bibliography

Primary Sources

Bulwer, Edward. *England and the English*. 2 vols. London: George Routledge and Sons, 1874 [1833].

Darwin, Charles. *Autobiography*, edited by Nora Barlow. New York: Norton, 1969.

– *On the Origin of Species*, edited by Joseph Carroll. Orchard Park: Broadview, 2003 [1859].

– *The Origin of Species*, 6th ed. New York: Mentor, 1958 [1872].

– *On the Origin of Species by Charles Darwin: A Variorum Text*, edited by Morse Peckham. Philadelphia: University of Pennsylvania Press, 1959.

– *The Life and Letters of Charles Darwin*, edited by Francis Darwin. 2 vols. New York: D. Appleton and Company, 1887.

– *The Correspondence of Charles Darwin*, vol. 2, *1837–1843*, edited by Frederick Burkhardt and Sydney Smith. Cambridge: Cambridge University Press, 1986.

– *The Correspondence of Charles Darwin*, vol. 3, *1844–1846*, edited by Frederick Burkhardt and Sydney Smith. Cambridge: Cambridge University Press, 1987.

– *The Correspondence of Charles Darwin*, vol. 5, *1851–1855*, edited by Frederick Burkhardt and Sydney Smith. Cambridge: Cambridge University Press, 1989.

– *The Correspondence of Charles Darwin*, vol. 6, *1856–1857*, edited by Frederick Burkhardt and Sydney Smith. Cambridge: Cambridge University Press, 1990.

– *The Correspondence of Charles Darwin*, vol. 7, *1858–1859*, edited by Frederick Burkhardt and Sydney Smith. Cambridge: Cambridge University Press, 1991.

– *The Correspondence of Charles Darwin*, vol. 8, *1860*, edited by Frederick Burkhardt, Duncan M. Porter, Janet Browne, and Marsha Richmond. Cambridge: Cambridge University Press, 1993.

– *The Correspondence of Charles Darwin*, vol. 10, *1862*, edited by Frederick Burkhardt, Joy Harvey, Duncan M. Porter, and Jonathan R. Topham. Cambridge: Cambridge University Press, 1997.

Darwin, Emma. *Emma Darwin: A Century of Family Letters 1792–1896*, edited by Henrietta Litchfield. 2 vols. London: John Murray, 1915.

'Degradation of Naval Surgeons.' *Lancet* 1 (1847): 680.

Draper, John William. *History of the Conflict between Science and Religion*. New York: D. Appleton and Company, 1897 [1874].

– *History of the Intellectual Development of Europe*. 2 vols. London: Bell and Daldy, 1864.

Evening Star. 2 July 1860.

Fawcett, Henry. 'A Popular Exposition of Mr. Darwin on the Origin of Species.' *Macmillan's Magazine* 3 (December 1860): 81–92.

Freemantle, W.H. *Charles Darwin, His Life Told*. 1892.

Gray, Asa. *Natural Selection Not Inconsistent with Natural Theology*. Boston: Ticknor and Fields, 1861.

Green, John. *Letters of John Richard Green*, edited by Leslie Stephen. London: Macmillan & Co., 1901.

The Guardian. 4 July 1860, 593.

Huxley, Leonard. *The Life and Letters of Sir Joseph Dalton Hooker*. 2 vols. London: John Murray, 1918.

– *Life and Letters of Thomas Henry Huxley*. 2 vols. New York: D. Appleton and Company, 1900.

Huxley, Thomas Henry. T.H. Huxley Papers. Imperial College, Huxley Archives.

– 'Bishop Wilberforce and Professor Huxley.' *The Times*. 1 December 1887, p. 8.

– *Darwiniana*. New York: D. Appleton and Company, 1897.

[–] 'The Origin of Species.' *Westminster Review* 17 (1860): 541–70.

– 'Science.' *Westminster Review* 61 (1854): 254–70.

– *Science and Hebrew Tradition*. London: Macmillan and Company, 1893.

– 'On the Theory of the Vertebrate Skull.' *Proceedings of the Royal Society of London* 9 (1857–59): 381–457.

John Bull. 6 July 1860, p. 422.

'Literature and Art.' *The Press*. 7 July 1860, p. 656.

Lyell, Charles. *Life, Letters, and Journals of Sir Charles Lyell*, edited by Mrs. Lyell. London: John Murray, 1881.

Marx, Karl, and Friedrich Engels. *The Communist Manifesto.* New York: Penguin, 1967 [1848].
'Monkeyana.' *Punch.* 18 May 1861, p. 206.
Newton, Alfred. 'Early Days of Darwinism.' *Macmillan's Magazine* 57 (February 1888): 241–9.
Owen, Richard. 'On the generation of the marsupial animals, with a description of the impregnated uterus of the kangaroo.' *Philosophical Transactions of the Royal Society of London* (1834): 333–64.
[–] 'Darwin *On the Origin of Species.*' *Edinburgh Review* 226 (April 1860): 487–532.
Powell, Baden. 'On the Study of the Evidences of Christianity.' In *Essays and Reviews: The 1860 Text and Its Readings*, edited by Victor Shea and William Whitla. 233–74. Charlottesville: University of Virginia Press, 2000.
'Section D.–Zoology and Botany, Including Physiology.' *The Athenaeum.* 7 July 1860, pp. 25–6.
'Section D.–Zoology and Botany, Including Physiology.' *The Athenaeum.* 14 July 1860, pp. 64–5.
Shelley, Mary Wollstonecraft. *Frankenstein; or, The Modern Prometheus*, edited by D.L. Macdonald and Kathleen Scherf, 2nd ed. Peterborough: Broadview, 1999.
[Sidgwick, Isabel.] 'A Grandmother's Tales.' *Macmillan's Magazine* 78 (1898): 425–35.
Stephen, Leslie. 'An Attempted Philosophy of History.' *Fortnightly Review* 27 (1880): 672–95.
Temple, Frederick. 'The Education of the World.' In *Essays and Reviews*, edited by Victor Shea and William Whitla. 137–80. Charlottesville: University of Virginia Press, 2000.
Tuckwell, W. *Reminiscences of Oxford*. London: Cassell and Company, 1901.
White, Andrew D. *History of the Warfare of Science with Theology in Christendom*. 2 vols. Buffalo: Prometheus, 1993 [1896].
Wilberforce, R.G. *Life of the Right Reverend Samuel Wilberforce,* vols. 2 and 3. London: John Murray, 1881.
– 'Professor Huxley and the Life and Letters of C. Darwin.' *The Times*, 29 November 1887, p. 10.
Wilberforce, R.G., and A.R. Ashwell. *Life of Samuel Wilberforce*, vol 2. London: John Murray, 1880.
Wilberforce, Samuel. MSS Wilberforce. Bodleian Library, Oxford University.
[–] Review of *On the Origin of Species. Quarterly Review* 108 (1860): 225–64.
[–] Review of *Essays and Reviews. Quarterly Review* 109 (1861): 248–305.

Secondary Sources

Altholz, Josef L. *Anatomy of a Controversy: The Debate over* Essays and Reviews, *1860–1864*. Aldershot: Scolar, 1994.

Aydon, Cyril. *Charles Darwin*. New York: Carroll and Graff, 2003.

Barton, Ruth. '"Huxley, Lubbock, and Half a Dozen Others": Professionals and Gentlemen in the Formation of the X Club.' *Isis* 89 (1998): 410–44.

Beatty, John. 'Speaking of Species: Darwin's Strategy.' In *The Darwinian Heritage*, edited by David Kohn. 265-282. Princeton: Princeton University Press, 1982.

Biagioli, Mario. *Galileo Courtier: The Practice of Science in the Culture of Absolutism* (Chicago: University of Chicago Press, 1993).

Bowler, Peter J. *Evolution: The History of an Idea*, rev. ed. Berkeley: University of California Press, 1989.

Brooke, John Hedley. 'Natural Theology and the Plurality of Worlds: Observations on the Brewster–Whewell Debate.' *Annals of Science* 34 (1977): 221–86.

– *Science and Religion: Some Historical Perspectives*. Cambridge: Cambridge University Press, 1991.

– 'The Wilberforce–Huxley Debate: Why Did It Happen?' *Science and Christian Belief* 13 (2001): 127–41.

Browne, Janet. *Charles Darwin*, vol. 2, *The Power of Place*. London: Pimlico, 2003.

– '"I Could Have Retched All Night": Charles Darwin and His Body.' In *Science Incarnate: Historical Embodiments of Natural Knowledge*, edited by Christopher Lawrence and Steven Shapin. 240–87. Chicago: University of Chicago Press, 1998.

Chapman, Matthew. *40 Days and 40 Nights: Darwin, Intelligent Design, God, OxyContin, and Other Oddities on Trial in Pennsylvania*. New York: Harper-Collins, 2007.

– 'God or Gorilla: A Darwin Descendant at the Dover Monkey Trial.' *Harper's Magazine* (February 2006): 54–63.

Cohen, I. Bernard. 'Three Notes on the Reception of Darwin's Ideas on Natural Selection (Henry Baker Tristram, Alfred Newton, Samuel Wilberforce).' In *The Darwinian Heritage*, edited by David Kohn. 589–607. Princeton: Princeton University Press, 1985.

Colp, Jr, Ralph. *To Be an Invalid: The Illness of Charles Darwin*. Chicago: University of Chicago Press, 1977.

– '"Confessing a Murder": Darwin's First Revelations about Transmutation.' *Isis* 77 (1986): 8–32.

– 'The Relationship of Charles Darwin to the Ideas of His Grandfather, Dr Erasmus Darwin.' *Biography* 9 (Winter 1986): 1–24.

Corsi, Pietro. *Science and Religion: Baden Powell and the Anglican Debate, 1800–1860*. Cambridge: Cambridge University Press, 1988.

Dawson, Gowan. *Darwin, Literature, and Victorian Respectability*. Cambridge: Cambridge University Press, 2007.

Desmond, Adrian. *Huxley: From Devil's Disciple to Evolution's High Priest*. London: Penguin, 1997.

– *The Politics of Evolution: Morphology, Medicine, and Reform in Radical London*. Chicago: University of Chicago Press, 1989.

Desmond, Adrian, and James Moore. *Darwin: The Life of a Tormented Evolutionist*. New York: Norton, 1994.

Eiseley, Loren. *Darwin's Century: Evolution and the Men Who Discovered It*. New York: Anchor, 1961.

Endersby, Jim. *Imperial Nature: Joseph Hooker and the Practices of Victorian Science*. Chicago: University of Chicago Press, 2008.

Feldhay, Rivka. *Galileo and the Church: Political Inquisition or Critical Dialogue?* Cambridge: Cambridge University Press, 1995.

Fox, Robert. 'The University Museum and Oxford Science, 1850–1880.' In *The History of the University of Oxford*, edited by M.G. Brock and M.C. Curthoys, vol. 6. Oxford: Clarendon, 1997.

Fyfe, Aileen. 'Conscientious Workmen or Booksellers' Hacks? The Professional Identities of Science Writers in the Mid-Nineteenth Century.' *Isis* 96 (2005): 192–223.

Gilley, Sheridan. 'The Huxley–Wilberforce Debate: A Reconsideration.' In *Religion and Humanism: Studies in Church History*, edited by Keith Robins, vol. 17. 325–40. Oxford: Blackwell, 1981.

Gilley, Sheridan, and Ann Loades. 'Thomas Henry Huxley: The War between Science and Religion.' *Journal of Religion* 61 (1981): 285–308.

Gould, Stephen Jay. *Bully for Brontosaurus: Reflections in Natural History*. New York: Norton, 1992.

– 'Impeaching a Self-Appointed Judge.' *Scientific American* 267 (July 1992): 118–21.

Hacking, Ian. 'Root and Branch.' *The Nation*, 8 October 2007, pp. 25–30.

Hitchens, Christopher. 'Equal Time: No Tax-Exempt Status for Churches That Refuse to Distribute Pro-Evolution Propaganda!' *Slate*, 23 August 2005.

Houghton, Walter E. *The Victorian Frame of Mind, 1830–1870*. New Haven: Yale University Press, 1957.

Hull, David. *Darwin and His Critics: The Reception of Darwin's Theory of Evolution by the Scientific Community*. Cambridge, MA: Harvard University Press, 1973.

James, Frank A.J.L. 'An "Open Clash between Science and the Church"? Wilberforce, Huxley, and Hooker on Darwin at the British Association, Oxford, 1860.' In *Science and Beliefs: From Natural Philosophy to Natural History*, edited by David Knight and Matthew Eddy. 171–93. Aldershot: Ashgate, 2005.

Jenson, J. Vernon. 'Return to the Wilberforce–Huxley Debate.' *British Journal for the History of Science* 21 (1988): 161–79.

Johnson, Curtis N. 'The Preface to Darwin's *Origin of Species*: The Curious History of the "Historical Sketch."' *Journal of the History of Biology* 40 (2007): 529–56.

Johnson, Paul. *The Offshore Islanders*. London: Weidenfeld and Nicolson, 1972.

Keynes, Randal. *Annie's Box: Charles Darwin, His Daughter, and Human Evolution* (London: Fourth Estate, 2001).

Kitcher, Philip. *Abusing Science: The Case against Creationism*. Cambridge: Cambridge University Press, 1982.

Kuhn, Thomas. *The Structure of Scientific Revolutions*, 2nd ed. Chicago: University of Chicago Press, 1970.

Larson, Edward J. *Summer for the Gods: The Scopes Trial and America's Continuing Debate over Science and Religion*. New York: Basic, 1997.

Levine, George. *Darwin and the Novelists: Patterns of Science in Victorian Fiction*. Chicago: University of Chicago Press, 1991.

– *Dying to Know: Scientific Epistemology and Narrative in Victorian England*. Chicago: University of Chicago Press, 2002.

Lightman, Bernard. 'Interpreting Agnosticism as a Nonconformist Sect: T.H. Huxley's "New Reformation."' In *Science and Dissent in England, 1688–1945*, ed. David M. Knight and Trevor H. Levere, 197–214. Aldershot: Ashgate, 2004.

– *The Origins of Agnosticism: Victorian Unbelief and the Limits of Knowledge*. Baltimore: Johns Hopkins University Press, 1987.

– '*Robert Elsmere* and the Agnostic Crisis of Faith.' In *Victorian Faith in Crisis: Essays on Continuity and Change in Nineteenth-Century Religious Belief*, edited by Richard J. Helmstadter and Bernard Lightman. 238–314. London: Macmillan, 1990.

– 'Science and Unbelief.' Conference on Science and Religion around the World, Green College, University of British Columbia, May 2007.

Livingstone, David N. 'Re-placing Darwinism and Christianity.' In *When Science and Christianity Meet*, edited by David C. Lindberg and Ronald L. Numbers. 183–202. Chicago: University of Chicago Press, 2003.

Lucas, John R. 'Wilberforce and Huxley: A Legendary Encounter.' *Historical Journal* 22 (1979): 313–30.

Macleod, Roy M. 'The X-Club: A Social Network of Science in Late-Victorian England.' *Notes and Records of the Royal Society of London* 24 (1970): 305–22.

Meacham, Standish. *Lord Bishop: The Life of Samuel Wilberforce 1805–1873.* Cambridge, MA: Harvard University Press, 1970.

Miller, David Philip. 'The "Sobel Effect": The Amazing Tale of How Multitudes of Popular Writers Pinched All the Best Stories in the History of Science and Became Rich and Famous While Historians Languished in Accustomed Poverty and Obscurity, and How This Transformed the World.' *Metascience* (2002): 185–200.

Moore, James. '1859 and All That: Remaking the Story of Evolution-and-Religion.' In *Charles Darwin, 1809-1882: A Centennial Commemorative*, edited by Roger G. Chapman and Cleveland T. Duval. 167–94. Wellington: Nova Pacifica, 1982.

– *The Post-Darwinian Controversies: A Study of the Protestant Struggle to Come to Terms with Darwin in Great Britain and America, 1870–1900*. New York: Cambridge University Press, 1979.

Morrell, Jack B., and Arnold W. Thackray. *Gentlemen of Science: Early Years of the British Association for the Advancement of Science*. Oxford: Oxford University Press, 1981.

Newsome, David. *The Victorian World Picture: Perceptions and Introspections in an Age of Change*. New Brunswick: Rutgers University Press, 1997.

Nichols, Peter. *Evolution's Captain: The Dark Fate of the Man Who Sailed Charles Darwin around the World*. New York: HarperCollins, 2003.

Novick, Peter. *That Noble Dream: The 'Objectivity Question' and the American Historical Profession*. Cambridge: Cambridge University Press, 1988.

Porter, Duncan M. 'On the Road to the *Origin* with Darwin, Hooker, and Gray.' *Journal of the History of Biology* 26 (Spring 1993): 1–38.

Rupke, Nicolaas A. *Richard Owen: Victorian Naturalist*. New Haven: Yale University Press, 1994.

Ruse, Michael. *The Darwinian Revolution: Science Red in Tooth and Claw*, 2nd ed. Chicago: University of Chicago Press, 1999.

– *The Evolution–Creation Struggle*. Cambridge, MA: Harvard University Press, 2005.

– *Monad to Man: The Concept of Progress in Evolutionary Biology*. Cambridge, MA: Harvard University Press, 1996.

Secord, James A. *Victorian Sensation: The Extraordinary Publication, Reception, and Secret Authorship of* Vestiges of the Natural History of Creation. Chicago: University of Chicago Press, 2000.

Stott, Rebecca. *Darwin and the Barnacle: The Story of One Tiny Creature and History's Most Spectacular Scientific Breakthrough*. London: Faber and Faber, 2004.

– 'Masculinities in Nineteenth-Century Science: Huxley, Darwin, Kingsley, and the Evolution of the Scientist.' *Studies in History and Philosophy of Biological and Biomedical Sciences* 35 (2004): 199–207.

Topham, Jonathan R. 'Beyond the "Common Context": The Production and Reading of the Bridgewater Treatises.' *Isis* 89 (1998): 233–62.

Turner, Frank. *Between Science and Religion: The Reaction to Scientific Naturalism in Late Victorian England*. New Haven: Yale University Press, 1974.

– 'The Victorian Conflict between Science and Religion: A Professional Dimension.' *Isis* 69 (1978): 356–76.

White, Paul. *Thomas Huxley: Making the 'Man of Science.'* Cambridge: Cambridge University Press, 2003.

Wollaston, A.F.R. *Life of Newton*. London: John Murray, 1921.

Yanni, Carla. *Nature's Museum: Victorian Science and the Architecture of Display*. London: Athlone, 1999.

Yeo, Richard. *Defining Science: William Whewell, Natural Knowledge, and Public Debate in Early Victorian Britain*. Cambridge: Cambridge University Press, 1993.

Index